OXFORD HI

THE ORIGINS OF MILITARY THOUGHT

from the Enlightenment to Clausewitz

AZAR GAT

CLARENDON PRESS · OXFORD

This book has been printed digitally and produced in a standard specification in order to ensure its continuing availability

OXFORD
UNIVERSITY PRESS

Great Clarendon Street, Oxford OX2 6DP

Oxford University Press is a department of the University of Oxford.
It furthers the University's objective of excellence in research, scholarship,
and education by publishing worldwide in

Oxford New York

Auckland Cape Town Dar es Salaam Hong Kong Karachi
Kuala Lumpur Madrid Melbourne Mexico City Nairobi
New Delhi Shanghai Taipei Toronto
With offices in
Argentina Austria Brazil Chile Czech Republic France Greece
Guatemala Hungary Italy Japan South Korea Poland Portugal
Singapore Switzerland Thailand Turkey Ukraine Vietnam

Oxford is a registered trade mark of Oxford University Press
in the UK and in certain other countries

Published in the United States
by Oxford University Press Inc., New York

Reprinted 2011

ISBN 978-0-19-820257-8

Printed and bound in Great Britain by CPI Antony Rowe,
Chippenham and Eastbourne

To Ruthie

ACKNOWLEDGEMENTS

THIS book is a revised and enlarged version of a doctoral thesis composed at St Antony's College, Oxford University, in 1984–6. It is my pleasure to thank my supervisor Sir Michael Howard, Regius Professor of Modern History, for his penetrating criticism, thoughtful advice, and warm encouragement during those two most rewarding years. I could not have hoped for more sympathetic and stimulating a teacher.

The depth of my intellectual debt to the works of Sir Isaiah Berlin is obvious. Professor Berlin was also so kind as to take the trouble to secure the funds needed for the completion of this work. I am particularly grateful to the Avi Foundation, the Edward Boyle Memorial Trust, and the British Council for their generous grants. Princeton University Press have kindly granted me permission to quote from their edition of Clausewitz's *On War*. The study of Montecuccoli was previously published in *War & Society*.

Finally, I would like to thank Mrs Roberta Rosen-Kerler whose devotion and good will in editing this work extended far beyond the demands of friendship.

A.G.

PREFACE

THE aim of this book is to show that the major currents of modern military thought emerged out of the cultural frameworks and the historical and philosophical outlooks of the Enlightenment on the one hand and the movement which may best be described as the Counter-Enlightenment or the German Movement on the other. The failure to recognize this culture-bound nature of military theory has often resulted in a narrow, unhistorical understanding of its development.

Although it dominated the eighteenth century and subsequently took a decisive role in shaping the terms in which military thinkers considered the theory of war, the very existence, not to mention the intellectual origins, vision, and scope of what I shall call the military school of the Enlightenment has not been recognized by modern commentators. Viewed through the highly polemic attitudes of Clausewitz and the German military school of the nineteenth century, the military thought of the eighteenth century has only been vaguely understood and labelled stereotypically as the 'geometrical military school'. Unwittingly, modern commentators have often simply reiterated Romantic rhetoric.

This problem has been equally damaging for the understanding of Clausewitz. His place in the framework of a general European reaction at the turn of the nineteenth century, against the ideas of the Enlightenment—a development which was particularly strong in Germany—has barely been recognized. This has been one of the main reasons for the fact that much of Clausewitz's work still remains unclear and puzzling, leaving room for endemically conflicting and unhistorical interpretations which continuously reflect the military and political convictions of each period, not least our own.

This book is not intended to cover all spheres and aspects of military thought in the period concerned. Such a task would not only exceed the space available in this volume but would also transgress my particular interests. First and foremost, this work is a historical and analytical account of the conceptions of military theory held by military thinkers in two successive—deeply philosophical—intellectual environments, which gave us the idea that there is or ought to be something called 'a theory of war'. However, that this topic involves a broad treatment of the major military developments and much of the doctrinal thinking of the period under discussion (in some cases more than in others) goes without saying.

Much the same applies to the emphases in the time-span covered by the book. The first two chapters, dealing with Machiavelli and Montecuccoli, are brief and introductory in nature, and focus on the decline of the classical conception of military theory and the earliest notable manifestation of a new one. In the main body of the book, covering the period from the middle of the eighteenth century to the 1830s, the disparity in scholarly attention has led me to adopt a slightly different approach in each of the two parts. In view of the paucity and polemic nature of the research on the military thinkers of the Enlightenment, I have focused in the first part on presenting an overall picture of their theoretical outlook against the background of the world-view of their time. On the other hand, the abundance of, and prevailing attitudes in, studies of Clausewitz have called for a more elaborate and critical approach in the second part.

CONTENTS

INTRODUCTION

Machiavelli and the Classical Notion of the Lessons of History in the Study of War

The idea that war could be studied systematically by historical observation, by the selection of successful forms of organization, and by the imitation of stratagems emerged in antiquity, and was powerfully revived—with a strong practical tendency—in the Renaissance. It was a counterpart to the tradition of classical political philosophy, the deductive conception of history, and the notion of a universal law of nature. Like them, it stemmed from historical experience in which fundamental change was hardly recognized and the basic features of human reality were perceived as enduring and recurring in numerous ways in differing periods and societies. Military theory was then simply a synthesis of the best military models of the known cultural past, whether in Greece or Rome. For Xenophon, in his *Hellenica* and *Anabasis*, the theory of war was comprised in the combat formation and drill of the phalanx, particularly the Spartan, while for Polybius in his *Histories* or for Vegetius in *De Re Militari* it consisted in the sophisticated organization and deployment of the Roman legion.

Roughly speaking, very little had changed from the classical era to Machiavelli's time in what can today be called the technological dimension of war, nor consequently in the character of war itself. The foot soldier, horse, armour, manual weapons, fortifications, and siege-machinery undeniably underwent considerable developments and transformations, and the importance of each fluctuated in a diversity of military establishments, the most prominent of which included the Persian, Greek, Macedonian, Gaul, Roman, Parthian, chivalrous, and Swiss models. Still, these weapon-systems remained remarkably similar, and the diversity of military models which were based upon them also revealed fundamental recurring characteristics. Historical experience thus offered an extensive testing ground of a relatively limited number of military systems, exposing their strong and weak points in multifarious circumstances.

In the Renaissance, Machiavelli attempted a synthesis of the whole of military experience from antiquity to the developments of the late Middle Ages. In this he brought the classical conception of the lessons of history to its pinnacle. His basic assumption was that despite historical change, man and society remained 'in essence' the same at all times and cultures because human nature was immutable: 'the world has always gone on in the same way,' he wrote; 'ancient kingdoms . . . differed from one another because of the difference in their customs, but the world remained the same'.[1] History could thus teach us lessons which were valid in every period. This conception, which dominated Machiavelli's political work, also guided his military thought.[2] But it was in the military sphere—rapidly and decisively influenced by technological change—that this outlook on history and theory faced an almost immediate breakdown.

The reason for Machiavelli's close attention to military affairs is obvious: he regarded the role of force as paramount both in domestic and foreign politics. Thus he discussed military affairs throughout his political works and later devoted to them a specialized study, *The Art of War* (1521). Here he sought to distil the lessons of military history and use them in devising a complete scheme for an army of his day.

The militia, the national army of citizens called to fight for their *patria*, was regarded by Machiavelli as the only proper form of military organization both from the social and the military points of view. This had been positively revealed in antiquity in the heyday of the Greek city-states and, more especially, of the Roman republic; in modern times this explained the extraordinary power of the small Swiss republic. The same lesson had been negatively demonstrated

[1] Niccolò Machiavelli, *The Discourses*, II, preface, in Allan Gilbert (ed.), *The Chief Works and Others* (Durham, 1965), i. 322.

[2] For the role of past lessons, among others in the military sphere, see e.g. *The Discourses*, I, pref., in *Chief Works*, i. For the widely discussed revival of the classical conception of history's purpose and lessons by the humanists and Machiavelli, see esp. Felix Gilbert, *Machiavelli and Guicciardini* (Princeton, 1965), chs. 4, 5; Denys Hay, *Annalists and Historians* (London, 1977), chs. 1, 5, 6, esp. pp. 93–4, and, for the case of Machiavelli, p. 113; Myron P. Gilmore, 'The Renaissance Conception of the Lessons of History', in his *Humanists and Jurists* (Cambridge Mass., 1963); Machiavelli is discussed on pp. 25–34.

in antiquity by the role played by mercenary armies in the decline of Greece and Rome. And in the modern period it was reaffirmed by the conduct of the disloyal, rapacious, treacherous, impudent, and cowardly *condottieri*, who were more dangerous to their employers than to the enemy, and who were responsible for the downfall in the international status of the once-proud Italian city-states. During his political career in the Chancellery of the Florentine republic and as the Secretary of the Office of Ten, Machiavelli witnessed the crippling effect of Florence's dependence on the *condottieri*, and was the driving force behind the attempt to re-establish the Florentine militia.[3]

As to the actual organization of the army, Machiavelli maintained that infantry armed with weapons for fighting at close quarters, protected by armour, and deployed in deep formation would break, under normal circumstances, the most vigorous cavalry charges. This lesson had been demonstrated numerous times by the Greeks and the Romans. With the decline of the Roman state and organized armies, it had been somewhat obscured; but it was strikingly redemonstrated—to the amazement of chivalrous Europe—with the revival of the classical formation of infantry by the Swiss on the battlefields of Burgundy and Italy. According to Machiavelli, infantry was therefore to be the backbone of a properly built army.[4]

Regarding the battle formation of the infantry itself, Machiavelli argued that, armed with sword and shield, and deployed in several flexible, manœuvrable, and mutually supporting squares, it would throw into disintegration and slay at close quarters enemy infantry armed with the pike and deployed in fewer, larger, and less manœuvrable squares. This had been demonstrated time and again in the great encounters of the Roman legion with the Macedonian and Seleucid phalanx—in Cynoscephalae, Magnesia, and Pydna—and had been analysed in depth by Polybius in a celebrated treatise in the *Histories* (XVIII, 28–32). The same lesson was corroborated—though perhaps less decisively—by the engagements between the

[3] See *The Prince*, ch. 12; and *The Art of War*, the principal theme of Bk. I. For the militia in Florentine and humanist tradition see C. C. Bayley's comprehensive *War and Society in Renaissance Florence: The* De Militia *of Leonardo Bruni* (Toronto, 1961); Machiavelli's involvement is described on pp. 240–315. For a much briefer summary of humanist attitudes see Quentin Skinner, *The Foundations of Modern Political Thought* (Cambridge, 1978) i. 76–7, 150–1, 173–5.

[4] *The Art of War*, II, in *Chief Works*, i. 602–4.

Spanish infantry armed with the sword and buckler, and the deep hedgehogs of the highly renowned Swiss infantry armed with the pike. Infantry, Machiavelli maintained, was therefore to be built by adopting the example of the Roman legion, though some of the features of the Macedonian and Swiss formations were also to be incorporated.[5]

As was the case in his political writings, Machiavelli freely adapted historical evidence to fit his argument. His hostility towards the *condottieri* made his account of their military conduct particularly tendentious. This has been exposed by modern research.[6] However, as pointed out by commentators, Machiavelli's aim was predominantly theoretical rather than historical.[7] Despite inaccuracies, he put forth a penetrating analysis of the principal military models of the past and achieved a remarkable synthesis of the legacy of classical military theory. Yet it took little time before his military views were struck by the full weight of an unprecedented historical change.

At the very time at which Machiavelli wrote and published *The Art of War*, the old forms of warfare were being revolutionized, predominantly because of the introduction of firearms. The Swiss formation could be regarded as a new Macedonian phalanx, while that of the Spaniards might have resembled in some respects the

[5] *The Art of War*, II and III, in *Chief Works*, vol. i, pp. 595–601 and 627–32 respectively.

[6] The reliability of Machiavelli's works as a source for the military events of his time was critically examined for the first time in Walter Hobohm's *Machiavellis Renaissance der Kriegskunst* (Berlin, 1913). For a concurrence with Hobohm and forceful defence of his work against criticism see Hans Delbrück, *History of the Art of War within the Framework of Political History*, (German original 1920; London 1985); iv. 101, 113; Delbrück's emphasis on understanding military affairs against their wider political background, and the legitimacy that he gave to limited strategy, 'the strategy of attrition', made him particularly aware of Machiavelli's bias against the *condottieri*. This bias has dominated the traditional historical view (for the interesting case of Clausewitz's attitude towards the *condottieri* as against his historicism, see below Ch. 7), and was still strongly expressed in Charles Oman, *The Art of War in the Middle Ages* (London, 1924), which reflected the 19th-cent. faith in the imperative of decision through battle; see esp. vol. ii, Bk. XII, ch. 2. But Machiavelli's account of *condottieri* warfare was convincingly criticized in Willibald Block, 'Die Condottieri: Studien über die sogenannten "unblutigen Schlachten"', *Historische Studien*, CX (1913); and recently the case for the *condottieri* was thoroughly made by Michael Mallett, *Mercenaries and their Masters: Warfare in Renaissance Italy* (London, 1974).

[7] See Delbrück, *History*, iv. 101, 113; and Felix Gilbert, 'Machiavelli: The Renaissance in the Art of War', in P. Paret (ed.), *Makers of Modern Strategy* (Princeton, 1986), 21–2.

Roman infantry. But the new arquebuses and guns could not be moulded into the old framework. The attempts to dismiss them as insignificant, or to adapt them into the paradigm of the classical battlefield as a 'new form' of archers and slingers,[8] were thwarted—the former in immediate failure, and the latter in decreasing achievements over a longer period of time—as their revolutionary effect on the battlefield grew ever stronger.

In the fictitious battle described in *The Art of War*, Machiavelli allowed the artillery to shoot only once and ineffectively before the armies closed. If commanders 'do rely on infantry and on the method aforesaid,' he wrote in *The Discourses*, 'artillery becomes wholly useless'.[9] Yet, while he was composing *The Discourses* and six years prior to the appearance of *The Art of War*, the guns of Francis I broke the dreadful Swiss infantry on the battlefield of Marignano (1515). And only a year after Machiavelli dismissed the significance of the new arquebuses, sarcastically remarking that they were useful mainly for terrorizing peasants,[10] the Spanish arquebusiers inflicted on the Swiss infantry its second great defeat at the Battle of Bicocca (1522).[11]

There have been some attempts to explain away and minimize Machiavelli's dismissal of firearms precisely when they were beginning to play an increasingly decisive role in the Italian wars of the late fifteenth and early sixteenth centuries. Felix Gilbert, for example, pointed out that Machiavelli's attitude to artillery in *The Discourses* (II, 17) was deliberately one-sided, having a polemic aim to restress the dominant role of valour.[12] Machiavelli's emphasis on moral forces is indeed undisputed, yet his attitude to firearms cannot be mainly understood as polemic tactics. This is certainly not the case with *The Art of War*, to which Gilbert does not refer in this

[8] *The Art of War*, II, in *Chief Works*, i. 597.

[9] Ibid. III, in *Chief Works*, i. 634; *The Discourses*, II, 17, in *Chief Works*, i. 371.

[10] *The Art of War*, II, in *Chief Works*, i. 625.

[11] For Renaissance warfare in the late 15th and early 16th cents. see Oman, *War in Middle Ages*; id., *The Art of War in the Sixteenth Century* (London, 1937); F. L. Taylor, *The Art of War in Italy 1494–1529* (Cambridge, 1921); Delbrück, *History*. For an updated narrative see J. R. Hale's chapters on military affairs in the *New Cambridge Modern History*, vols. i–iii (Cambridge, 1957, 1958, 1968). On attitudes to firearms see id., 'Gunpowder and the Renaissance', in his *Renaissance War Studies* (London, 1983); this comprehensive article surprisingly does not deal with Machiavelli. See also id., *War and Society in Renaissance Europe* (London, 1985).

[12] Gilbert, 'Machiavelli: The Renaissance in the Art of War', in E. Earle (ed.), *Makers of Modern Strategy* (Princeton, 1943), 14–15.

context. *The Art of War* is Machiavelli's positive and complete scheme for the building of armies, and reflects the full scope of his military outlook. His ideal army is totally of Roman and Macedonian-Swiss form, and though artillery is introduced, its significance and role in battle could not have been more belittled.[13]

Most commentators, however, have been critical of Machiavelli's military ideas. Oman, for example, wrote that Machiavelli, though very perceptive, was mistaken in all his major predictions of future military developments, particularly regarding firearms.[14] How then did the Florentine thinker, famous for his penetrating and sobering insights into the complexity of human relations, politics, and society, fail to recognize one of the most important milestones of military history? Clausewitz, otherwise an admirer of Machiavelli, pointed to the obvious reason in a letter to Fichte:

> So far as Machiavelli's book on the art of war itself is concerned, I recall missing the free, independent judgment that so strongly distinguishes his political writings. The art of war of the ancients attracted him too much, not only its spirit, but also in all of its forms.[15]

This line of explanation—independently arrived at by later commentators[16]—is undoubtedly true, but should be expanded. As mentioned above, the reasons for Machiavelli's great misjudgement go deeper. It can only be understood in the context of his conceptions of history and theory. His way of thinking in attempting to overcome the challenge of artillery is most revealing. He sought an analogy in antiquity:

> In approaching the enemy, infantry can with greater ease escape the discharge of artillery than in Antiquity they could escape the rush of elephants or of scythed chariots and of other strange weapons that the Roman infantry had to oppose. Against these they always found a remedy. And so much the more easily they would have found one against artillery.[17]

Machiavelli could not accept firearms as a significant military and political innovation because this would have undermined not only

[13] *The Art of War*, in *Chief Works*, i. 632.

[14] Oman, *War in Sixteenth Century*, pp. 93–4.

[15] Letter to Fichte, 11 Jan. 1809, in W. M. Schering (ed.), *Clausewitz, Geist und Tat* (Stuttgart, 1941), 76; P. Paret, *Clausewitz and the State* (Oxford, 1976), 176. See extensively below in the chapters on Clausewitz.

[16] Cf. Oman, *War in Middle Ages*, ii. 311.

[17] *The Discourses*, II. 17, in *Chief Works*, i. 371. For a similar reasoning see also *The Art of War*, III, in *Chief Works*, i. 637.

his model for military organization and virtues—the Roman army—but also the foundations of his historical and theoretical outlook. Such acceptance would have implied a historically unprecedented, fundamental change in the well-known recurring patterns of past warfare, invalidating the lessons offered by the historical perspective of two thousand years.[18]

The classical legacy continued to form the intellectual background and source of historical reference for military thinking—among other spheres of European culture—until the end of the eighteenth century. The works of the classical authors were widely studied and considered the best material for military instruction. These included the histories of Herodotus, Thucydides, Livy, Tacitus, Plutarch, and particularly of those historians who emphasized military aspects, such as Xenophon, Polybius, and Caesar. Equally popular were the military treatises of such authors as Arrian, Vegetius, Frontinus, Aelian, Polyaen, Vitruvius and the Byzantine emperors Maurice and Leo. They were published in numerous editions and continually elicited vivid attention and extensive commentary between the late fifteenth and late eighteenth centuries.[19]

Initially, the classical military models, when synthesized with modern firearms, were still of great relevance and influence. Machiavelli, and later the celebrated classical scholar and humanist philosopher, Justus Lipsius, in his *Politicorum libri sex* (1589) and *De militia Romana* (1596), propagated the organization, discipline, and flexible internal division of the Roman legion. These inspired the military reforms associated with Maurice of Orange and his Nassau cousins during the Dutch wars of independence, and the organization of the Swedish army under Eric and Gustavus Adolphus.[20] However, the old weapons and formations were gradually being abandoned. The pike, the last notable remnant of

[18] For a similar criticism of Machiavelli's political thought cf. H. Butterfield, *The Statecraft of Machiavelli* (London, 1955).

[19] For a comprehensive survey of the reprints of the military works of antiquity during this period see Max Jähns, *Geschichte der Kriegswissenschaften* (Munich and Leipzig, 1889), 244–8, 447–54, 869–75, 1142–3, 1461–3, 1823–37.

[20] See esp. W. Hahlweg, *Die Heeresreform der Oranier und die Antike* (Berlin, 1941); M. Roberts, *Gustavus Adolphus* (London, 1958), vol. ii, ch. III; G. Oestreich, *Neostoicism and the Early Modern State* (Cambridge, 1982), ch. 5, 'The Military Renascence'; and G. Rothenberg, 'The Seventeenth Century', in Paret (ed.), *Makers of Modern Strategy*, pp. 32–63.

ancient and medieval warfare, went out of use by the end of the seventeenth century when it was replaced by the bayonet fixed to the muzzle of the musket. This, together with the growing effectiveness of the musket and field-gun, led to a decrease in the depth of battle formation throughout the eighteenth century. The line won the day.[21] No longer did the classical military legacy represent a homogeneous historical experience or provide direct analogies and lessons for the present as Machiavelli had assumed in *The Art of War*.

Yet, the emergence of a strong opposition to the linear formation in the eighteenth century went hand in hand with a powerful revival in the reference to, and interest in, ancient warfare and military works. Folard advocated the restoration of the shock effect of the pike and the column in his *Histoire de Polybe* (1724–30), and in the 1770s his disciple Mesnil-Durand sparked the great doctrinal controversy between the *ordre profond* and the *ordre mince*. This led to a compromise and to the introduction of the famous column of the Wars of the Revolution and Napoleon as a formation for manœuvres and fighting.[22] As we shall see, de Saxe, Puységur, Guichard, Turpin, and Maizeroy relied on the ancient models and authorities almost as heavily as Machiavelli. There was almost no military thinker in the Enlightenment who did not refer to antiquity to some extent. Even the characteristic debate of the seventeenth and eighteenth centuries, as to whether the ancients or the moderns were culturally superior, was not lacking in the military sphere.

The notion of a fundamental historical change began to emerge with the Enlightenment, but a new attitude to the past, including military history, took shape only at the close of the eighteenth century. First, after the French Revolution, Tempelhoff, Bülow, and Clausewitz observed a new, 'modern' experience. In a direct reaction against the military thinkers of the French Enlightenment, Tempelhoff wrote that theory had to be based on contemporary experience rather than on the history of the Greeks and Romans.[23] Still more important was the emergence of historicism with its

[21] For the European armies in this period of transition see D. Chandler, *The Art of Warfare in the Age of Marlborough* (London, 1976).

[22] For a fuller account see below, ch. 2. For the doctrinal controversy see Jean L. A. Colin, *L'Infanterie au* XVIII*e* siècle (Paris, 1907); Robert S. Quimby, *The Background of Napoleonic Warfare* (New York, 1957).

[23] G. Tempelhoff, *History of the Seven Years War* (London, 1793), i. 84.

supreme sensitivity to the diversity of historical experience and the uniqueness of every period. Clausewitz, who introduced the historicist outlook into military thought, wrote:

> Wars that bear a considerable resemblance to those of the present day, especially with respect to armament, are primarily campaigns beginning with the War of the Austrian Succession. Even though many major and minor circumstances have changed considerably, these are close enough to modern warfare to be instructive . . . The further back you go, the less useful military history becomes . . . The history of Antiquity is without doubt the most useless . . . We are in no position . . . to apply [it] to the wholly different means we use today.[24]

The relative uniformity of historical experience as the basis for a theory of war which could be derived by direct observation, analysis, and critical analogy from the major military models of the past, and applied to the similar conditions of the present, was therefore gradually breaking down in the early modern period. Yet, this development was more than matched by the growth of a powerful, new theoretical ideal to subject all spheres of reality, including war, to the rule of reason. This ideal was greatly stimulated by the vision and achievements of the natural sciences which also put forward a new systematical model: to reveal the universal principles that dominate the diversity of phenomena. The overwhelming success of this enterprise, culminating in Newtonian science, was one of the principal driving forces of the Enlightenment and generated a corresponding awakening of military thought. But the proto-scientific outlook had already been influencing military theory in the seventeenth century.

[24] Carl von Clausewitz, *On War*, M. Howard and P. Paret (eds.) (Princeton, 1976), Bk II, ch. 6, p. 173.

Part One

THE MILITARY SCHOOL OF THE ENLIGHTENMENT

1

Montecuccoli

The Impact of Proto-Science on Military Theory

Known today only to a small circle of scholars, Raimondo Montecuccoli (1609–80) was regarded in the eighteenth century—much as Clausewitz has been in the last two centuries—as the most distinguished modern military thinker, whose widely cited and highly influential work was a classic that offered the foundations of a general theory of war.

What, then, was Montecuccoli's theoretical outlook, and, inseparably, what were its origins? While his life story and military career have had their normal share of historiographical attention, Montecuccoli's intellectual world—despite the evidence and despite his reputation as a 'military intellectual'—has not been explored nor connected to his military thought. The following chapter is therefore merely an introduction to a much-needed extensive study.

Deeply involved in the great intellectual fermentation in the first half of the seventeenth century, Montecuccoli gave expression to the ideas and attitudes of the late humanists, and was an enthusiastic student of the powerful tradition of research into the occult, alchemy, and natural magic. This was one of the major trends of the evolving scientific enterprise of the sixteenth and seventeenth centuries, and was widespread among the élite circles of the Habsburg empire.

Montecuccoli was born in 1609 to a noble family from the vicinity of Modena in northern Italy. Entering the Imperial army, he saw active service throughout the Thirty Years War, rising to the rank of general and distinguishing himself as a cavalry leader. After the war, he carried out diplomatic missions and commanded the Imperial forces in the Nordic war in Poland. In 1664 he defeated the invading Turkish army in his greatest battle at St Gotthard, and in 1673, during the Dutch war of Louis XIV, he conducted his most celebrated campaign against Turenne on the Rhine which was to be admired throughout the eighteenth century as a model of manœuvre. He again faced Turenne in the same theatre of operations in the campaign of 1675. Promoted to the rank of field marshal, Montecuccoli was appointed President of the Imperial War Council in 1668, and in this capacity he took the first steps in creating a professional standing army. Though awarded the titles Prince of the Empire and Duke of Melfi in 1679, his last years were clouded by power struggles and professional disputes with rivals both in court and within the army.[1]

We know very little about Montecuccoli's early education, but as a general he is described by a contemporary as a formidable intellectual figure with an extraordinary range of interests—well known as such even in the highly cultural environment of Vienna—and a patron of the sciences who possessed a huge library.[2] His intense intellectual preoccupations are clearly revealed in his major works, composed during three lulls in his military career.[3]

There has survived the varied list of sources—forty-five in all—that Montecuccoli used for the writing of his first major work, the *Treatise on War* (*Trattato della guerra*), composed while he spent four years in Swedish captivity in Stettin (1639–43).[4] The *Zibaldone*, Montecuccoli's extensive reference work, composed during the post-Westphalian period (1648–54), has also been preserved and includes sixty-nine bibliographical items.[5]

[1] The standard biographies of Montecuccoli are still Cesare Campori, *Raimondo Montecuccoli, la sua famiglia e i suoi tempi* (Florence, 1876); and Tommaso Sandonnini, *Il Generale Raimondo Montecuccoli e la sua famiglia* (Modena, 1914). Many articles and dissertations have been written in the last two centuries on Montecuccoli's military career; for a modern and comprehensive study of his greatest battle, see Georg Wagner, *Das Türkenjahr 1664, Raimund Montecuccoli, die Schlacht von St. Gotthard-Mogarsdorf*, issue 48 of *Burgenländische Forschungen* (1964). Two recent works are Hans Kaufmann, 'Raimondo Graf Montecuccoli 1609–1680', (doct. diss.; Berlin, 1972); and Thomas Barker, *The Military Intellectual and Battle: Montecuccoli and the Thirty Years War* (New York, 1975). For a concise overall account see Gunther Rothenberg, 'The Seventeenth Century' in Paret (ed.), *Makers of Modern Strategy*, pp. 55–63.

[2] The Italian tourist Abbé Pacichelli, cited by E. Vehse, *Memoirs of the Court and Aristocracy of Austria* (London, 1856), i. 432–4; Montecuccoli's character is depicted as cold, unsympathetic, and intriguing.

[3] Piero Pieri, 'La formazione dottrinale di Raimondo Montecuccoli', in *Revue internationale d'histoire militaire*, III (1951), 92–125.

[4] This list of sources has not been printed. The authors included are cited by Barker from the manuscript in Montecuccoli's family archive in Modena; see *Montecuccoli*, p. 227.

[5] This work has not been printed either; the authors and works are cited in A. Veltzé (ed.), *Ausgewählte Schriften des Raimund Fürsten Montecuccoli* (Vienna, 1899), vol. i, pp. cxiii–cxx.

Montecuccoli's central work from this period is *On the Art of War* (*Del arte militare*), a concise version of the *Treatise*, laying special emphasis on fortifications and siegecraft. The references in the last version, the celebrated *On the War against the Turks in Hungary* or *Aphorisms* (*Della guerra col Turco in Ungheria*; 1670)—a demonstration of Montecuccoli's military ideas through his campaign against the Turks—provide another insight into his intellectual background.[6]

These source lists, reference works, and military writings reveal a remarkable continuity in Montecuccoli's interests and ideas. They indicate that the thirty-two-year-old colonel and author of the *Treatise on War* had already consolidated his theoretical outlook and military conceptions, which underwent no further fundamental changes. Indeed, the *Treatise* is the largest of Montecuccoli's works, and since none of them were published during his lifetime, it is perhaps only accidental that the *Treatise* remained unpublished when his two later military discourses appeared at the beginning of the seventeenth century.

In the source list to his first work, Montecuccoli cites extensively the classical and the contemporary military authors whose full influence on his work has yet to be studied.[7] The conceptual framework of his work was undoubtedly influenced by systematic military treatises such as Giorgio Basta's *Il maestro di campo generale* (1606), Henri de Rohan's *Le Parfait Capitaine* (1631), Wallhausen's *Corpus militare* (1617), and perhaps also Mario Savorgnano's *Arte militare terrestre e maritima* (1599; not cited by Montecuccoli). However, it is mostly in the scope of general works and non-military authors cited by Montecuccoli that the clue to his outstanding theoretical outlook and endeavour is to be sought.

[6] Montecuccoli's extensive writings have been compiled from the Vienna War Archives in Veltzé's four vol. edn. The first and second vols. contain Montecuccoli's military works, the third his historical writings, and the fourth correspondence and miscellaneous works. The *Treatise on War* has not been published elsewhere. *On The Art of War* and esp. *The War against the Turks in Hungary* were published in all the major European languages with the exception of English. Only one of Montecuccoli's smaller works, *On Battle* (*Della battaglie*), appeared in English in Barker, *Montecuccoli*.

[7] A short account of these authors, based on Jähns, is given by Barker, *Montecuccoli*, pp. 55–8, 227, who also cites the intellectual authorities with little understanding and some factual errors.

These authorities fall into two major categories, the first being political authors and essayists. Montecuccoli cites Machiavelli's writings, Campanella's *Monarchia Hispania*, Bacon's *Essays*, and many other then very popular and now almost forgotten works such as those of the French man of letters, J. L. Guez de Balzac (1597–1654), particularly his *Le Prince*.[8] Yet the dominant influence on his work was that of the late humanist tradition as propounded by Justus Lipsius (1547–1606).

Lipsius's intellectual influence throughout Europe and in the Habsburg Empire in the late sixteenth and early seventeenth centuries was outstanding.[9] It was equally unparalleled in the military field, where Lipsius was the major proponent of Roman military values and practices, and influenced Maurice of Nassau's military reforms.[10] His principal influence on Montecuccoli's military works was, however, quite different. In Lipsius's celebrated *Six Books of Politics* (1589), which reflected the increasing dominance of the centralized state, Montecuccoli found a comprehensive and systematic presentation of war within a political framework, derived from political motives and directed towards political aims. As we shall see, Book I of the *Treatise on War*, Montecuccoli's earliest theoretical work, directly refers to, and closely follows Lipsius's conceptions.[11]

Similar attitudes to war were offered by Aristotle and by the Roman stoics—the school most popular among the humanists—particularly Cicero and Seneca. And they were also central to the jurist tradition. Cicero and Grotius are cited in this connection in the opening of *On the War against the Turks in Hungary*.[12]

[8] Machiavelli is cited as no. 13 in the *Zibaldone*, in which the first twenty items are, broadly speaking, political. *Monarchia Hispania* by Campanella, one of Montecuccoli's favourite authors who had also appeared in the earlier source list, is no. 17. Bacon's *Essays* comprise no. 2; Balzac's works are cited as nos. 5 and 18.

[9] See Oestreich's excellent *Neostoicism and the Early Modern State*; J. L. Saunders, *Justus Lipsius, The Philosophy of Renaissance Stoicism* (New York, 1955); and for Lipsius's popularity and influence in the Habsburg empire, R. J. W. Evans's highly comprehensive studies: *Rudolf II and his World, A Study in Intellectual History 1576–1612* (Oxford, 1973), esp. pp. 95–6, and, in the context of humanist culture, 116–61; id., *The Making of the Habsburg Monarchy 1550–1700* (Oxford, 1979), esp. pp. 25, 113.

[10] See Introd. n. 20 above.

[11] Oestreich, *Neostoicism*, pp. 80–1; unaware of Montecuccoli's scientific interests, Oestreich is mistaken, however, in attributing Montecuccoli's conception of science to Lipsius; see Montecuccoli's list of sources to the *Treatise*, and nos. 3 and 4 of the *Zibaldone*.

[12] For references to Aristotle, Cicero, and Seneca see mainly the introds. to *On the Art of War*, and *The War against the Turks* and, for Aristotle, see also *Zibaldone*, no. 6.

The second and even more extensive category of authors and works cited by Montecuccoli is truly remarkable. More than half of the *Zibaldone*, comprising about forty works, is a compendium of the great authorities of the occult and magical natural philosophy, covering Paracelsian alchemy and medicine, and Hermetic, cabbalistic, and Rosicrucian wisdom and visions. Indeed, some of these authorities were previously included in the source list to the *Treatise*.

One of those cited in the *Treatise* is the English Hermetic, cabbalistic, and Paracelsian natural philosopher, Robert Fludd (1574–1637), famous throughout Europe for his works on mathematical mystery and magic which he defended in a celebrated debate against Kepler. According to a contemporary, Montecuccoli 'was able to recite [his works] word for word'.[13] Another occultist cited in the source list is Johann Faulhaber, the author of *Magia arcana Coelestis sive Cabalisticus* (1613).[14] Johann Amos Comenius (1592–1670), the influential Czech bishop and philosopher of education, whose pacific universal, and humanist ideas were deeply embedded in the mystical tradition, is also included.[15] Finally, the appearance in the source list of the works of Georgius Agricola (1494–1555), one of the pioneers of modern geology and mineralogy, also attest to Montecuccoli's keen interest in the scientific thought of the time.[16]

This interest is fully revealed in the *Zibaldone*. Tommaso Campanella (1568–1639), the celebrated mystical and millenarian natural philosopher and political thinker mentioned earlier in the political section, is represented by an additional seventeen works, covering the full range of his metaphysical thought (nos. 22–38). Another prominent representative of the Italian occultist natural philosophy is Gianbattista Porta, whose popular *Magia Naturalis*

[13] Cited by Vehse, *Memoirs*, i. 434; Evans, *Habsburg Monarchy*, pp. 347–8. For Fludd, see Allen G. Debus, *The Chemical Philosophy, Paracelsian Science and Medicine in the Sixteenth and Seventeenth Centuries* (New York, 1977), i. 205–93; and F. A. Yates, *Giordano Bruno and the Hermetic Tradition* (London, 1964), ch. XXII; id., *The Art of Memory* (London, 1966), ch. XV, and *Theatre of the World* (London, 1969), chs. III–IV.

[14] See Evans, *Habsburg Monarchy*, p. 397.

[15] Ibid., esp. pp. 395, 399; and Evans, *Rudolf II*, esp. pp. 82, 276–7, 283–5, 290.

[16] Frank D. Adams, *The Birth and Development of the Geological Sciences* (London, 1938), ch. VI.

is cited (no. 40; 1644 edn.).[17] Among the founders of the Hermetic tradition, the famous Raymon Lull (1236–*c*.1316), the first to introduce the secret calculations of Jewish cabbala into European thought, is represented by his equally famous *Secreti di Natura*, translated together with St Albertus Magnus's *Delle cose minerali e metalliche* (no. 62; 1557 edn.).[18]

Extensive reference is made to the most important physicists and chemist-alchemists of the period, including Valerianus Magnus (1585–1661), the student of the vacuum (no. 39); Johann Rudolph Glauber (1603/4–1668/70), chemist and physicist (nos. 41–2); Andreas Libavius, the anti-Paracelsian chemist (no. 51; many works are cited); Zacharias Brendel (1592–1638), MD chemist and alchemist (no. 53); Johann Hartmann (1568–1631), the first professor of chemistry in Europe, holding a chair in Marburg (no. 55); and Oswald Croll, Paracelsian chemist and physician, who wandered through Europe finding an audience for his secret teaching in the Habsburg provinces and court, and whose *Basilica Chymica*, cited by Montecuccoli (no. 57), was published in eighteen editions between 1609–58.[19]

This group is inseparable from the corpus of medical works listed in the *Zibaldone*, most of which are by Paracelsian authors, including Johann Schröder (1600–64), author of the widely read *Pharmacopoeia Medico-chymica* (no. 45); Pierre Jean Fabre (d. 1650), graduate of Montpellier and author of the equally popular *Palladium Spagyrica* (no. 48); Lazar Riverius, another representative of and well-known professor at Montpellier (no. 58); J. B. van Helmont (1579–1644), a medical doctor of European renown (no. 50); and Jean Béguin, author of the pharmacological *Tyrocinium*

[17] See D. P. Walker, *Spiritual and Demonic Magic from Ficino to Campanella* (London, 1958); and for the Italian nature philosophers, J. H. Randall Jun., *The Career of Philosophy from the Middle Ages to the Enlightenment* (New York, 1962), 197–220.

[18] For Lull and Lullism, see J. N. Hillgarth, *Raymon Lull and Lullism* (Oxford, 1971); and F. A. Yates, *The Occult Philosophy in the Elizabethan Age* (London, 1979), ch. I.

[19] For Magnus's science and mysticism, see Evans, *Habsburg Monarchy*, esp. pp. 330, 337, 342, and for Glauber's influence in the Viennese court pp. 361 and 365. For Libavius see B. Easlea, *Witch-hunting, Magic and the New Philosophy* (Sussex, 1980), 107, and Debus, *Chemical Philosophy*, i. 169–73. For Hartmann and Croll, see ibid., pp. 125 and 117–24 respectively. For all the authorities cited, see also vols. vii and viii of L. Thorndike's *magnum opus*, *A History of Magic and Experimental Science* (New York, 1923–58).

Chymicum which appeared in no less than forty-one editions between 1610 and 1690 (no. 52; 1640 edn. is cited).[20]

Finally, still very closely related, is the popular literature of the various secret and traditional prescriptions. An example of this is the *Secreti* (1561) of Signora Isabella Cortese (no. 61; 1603 edn. is cited).[21]

Not all of the authors and works cited above were Paracelsian. Some, notably Libavius, were even opponents of the magical tradition. So also was Francis Bacon whose *Essays* are cited twice in the *Zibaldone*, both in the philosophical and the scientific sections.[22] While he rejected the mechanical-mathematical philosophy, and was a true child of the experimental enterprise attempting to control nature by discovering its secrets, Bacon was also one of the well-known critics of natural magic. Yet, the overwhelming majority of the natural philosophers cited both in the source list to the *Treatise on War* and in *Zibaldone* are Paracelsian, and they leave little doubt where Montecuccoli's interests lay.

Where it was noticed, this fact caused some concern among Montecuccoli's interpreters about the 'scientific' nature of his outlook.[23] This concern is obviously somewhat anachronistic and tends to assume a standard concept of science, as perceived by the men of the eighteenth century. Indeed, with the triumph of the mechanical-mathematical interpretation of nature, the occult tradition of natural philosophy was expelled from the domain of science as superstition. However, until the late seventeenth century, the struggle between the contending currents of natural philosophy still raged, and Newton's secret research into the occult was the last remarkable example of this. Montecuccoli's interests reflected the enormous, sometimes passionate interest of the educated social and political élite throughout the Habsburg empire in all spheres of the occult and natural magic. His title as the Protector of the *Leopoldinische Akademie der Naturforscher des heiligen Römischen*

[20] For Schröder, see Thorndike, *Magic and Experimental Science*, viii. 88–92. For Fabre and Béguin, see Debus, *Chemical Philosophy*, pp. 261 and 167–8 respectively. For Helmont and his Paracelsianism, ibid. 295–343, and W. Pagel, *Joan Baptista van Helmont* (Cambridge, 1982). For Riverius see the latter, pp. 37–62.

[21] Thorndike, *Magic and Experimental Science*, vi. 218.

[22] Nos. 2 and 40; the first Latin edn. of the *Essays* is cited, 'Sermones fideles', Lugd. Batav. 1641; see R. W. Gibson, *Bacon: A Bibliography of his Works and Baconiana to the Year 1750* (Oxford, 1950).

[23] Barker, *Montecuccoli*, pp. 5, 58.

Reiches (which was later to move to Halle) was typical of the patronage that the court and magnates bestowed upon the great proponents of these arts.[24]

It was thus from within this highly involved intellectual environment, in the context of the scholarly tradition of the late Renaissance and the extensive proto-scientific awakening, that Montecuccoli set out to undertake a scientific study of war.

In the opening of his earliest work Montecuccoli wrote:

Many ancients and moderns have written on war. Most of them, however, have not crossed the boundaries of theory. When some, such as Basta, Melzi, Rohan, la Noue, etc. have combined theory with its application, they have either undertaken to cultivate only one part of this vast field, or have restricted themselves to generalities, without getting down to the details of the supporting sciences . . . which make the perfect military general. It is impossible to understand the whole fully, if one is not familiar with its constitutive parts.[25]

Thus, he wrote in his second, more concise work, 'I have thought, in a limited framework, to summarize methodically the exceedingly vast territory of this science' which deals with an art of the utmost political importance.[26]

Like all sciences, the science of war aims to reduce experience to universal and fundamental rules.[27] These can then be applied to particular times and circumstances by means of skilful judgement, which is necessary in order to put the individual examples in a general perspective.[28]

In his celebrated *On the War against the Turks in Hungary*, written late in life, Montecuccoli offered a sophisticated epistemological account of this process, directly referring to, and closely following, Aristotle's analysis in the *Metaphysics* (A.I.):

The innate force of reason, while comprehending the objects, also turns them into concepts which it stores in the memory. From several combined

[24] For Montecuccoli and the Academy, see Veltzé (ed.), *Ausgewählte Schriften (AS)*, vol. i, pp. cxxx. For the occult culture in the Habsburg empire and court, including the contents of many libraries and reading lists which are very similar to Montecuccoli's, see again Evans, *Rudolf II*, esp. ch. 6, and id., *Habsburg Monarchy*, esp. ch. 10.

[25] 'An den Leser', *Abhandlung über den Krieg*, in Veltzé (ed.), *AS* i. 5–6.

[26] Introd. to *Von der Kriegskunst*, *AS* ii. 29; Cicero is cited on the importance of the art of war.

[27] Ibid.; see also *Abhandlung über den Krieg*, *AS* i. 7.

[28] *Abhandlung über den Krieg*, *AS* i. 7–8; citing the 'physicist' as an illustration.

recollections, experience emerges, and from many experiences there springs general understanding, which is the beginning of all sciences and arts.[29]

Hence the intimate relationship and interdependence between the theory of general rules and practice. While theory is derived from reality, it then serves to guide and judge action. Each is essential to the other.[30] Thus 'as the mathematician uses to do', the first part of Montecuccoli's book offers the principles of the art of war, while the second applies them—'like derivatives'—to the war against the Turks in Hungary.[31]

The universal rules of war encompass 'the whole of world history from the beginning of things'. There is 'no remarkable military deed . . . [that] cannot be reduced to these instructions'.[32] 'Disregarding the invention of artillery, which has somewhat changed the forms of war, the rest of the rules remain correct and valid.'[33]

Book I of the *Treatise on War* is an extensive study of the nature and political context of war, based on Lipsius's discussion in Books 5–6 of the *Six Books of Politics* to which constant reference is made. Wars are divided into internal and external, and their causes are elaborated under the headings of either remote or immediate. The prerequisites of just wars are discussed. The political preparations for war, particularly the striking of alliances, are described as well as the preparations of military means, divided into provisions, arms, and money. Lastly, the army itself is examined, including the hierarchy of command, and—reflecting neo-classical notions—recruitment methods; native soldiers are declared to be greatly superior to foreign troops. Book II deals with the conduct of war, and the final Book III with the conclusion of war and the attainment of a favourable peace, which was the purpose of the war.[34]

[29] *Vom Kriege mit den Türken*, *AS* ii. 199.

[30] Ibid.

[31] Ibid. 200.

[32] *Von der Kriegskunst*, *AS* ii. 29.

[33] *Vom Kriege mit den Türken*, *AS* ii. 200.

[34] For the three stages of war, see J. Lipsius, *Sixe Books of Politickes* (London, 1594), Bk. V, ch. 3; for the two types and the causes of war, ibid. and VI, 2–3, Montecuccoli, *Abhandlung*, *AS* i. 21–4; on just war, Lipsius, V, 4, Montecuccoli, i. 24–5; on the three types of military means, Lipsius, V, 6, Montecuccoli, i. 75–6; on military command, Lipsius, V, 14–17, Montecuccoli, i. 81–92; on recruitment, in the footsteps of Vegetius and Machiavelli, Lipsius, V, 9–12, Montecuccoli, i. 95; and on the favourable peace as the aim of war, Lipsius, V, 18–20. See also Oestreich, *Neostoicism*, pp. 80–1.

In Montecuccoli's later works these themes are repeated in a much more concise form, some of which are compressed into aphorisms and tables. Referring to Grotius's citation of Cicero's definition of war as 'a conflict with the use of violence', Montecuccoli wrote: 'War is an activity to inflict damage in every way; its aim is victory.'[35] The art of war is defined as 'the study of the good manner to conduct war'.[36] Wars are divided according to location, whether on land or at sea; type, whether defensive or offensive; and circumstance, whether civil or internal, just or unjust.[37]

Montecuccoli's extensive and systematic treatment of military organization and the conduct of war, encompassing training, discipline, supply, intelligence, fortifications, marching, encamping, and fighting, clearly reveals the tension between historical change on the one hand and Montecuccoli's theoretical ideals and historical notions on the other. His universal science of war was obviously simply a reflection of the warfare of his day, between the campaigns of Gustavus Adolphus and the early wars of Louis XIV.

At the political and social level, this was primarily linked with the rise of the centralized state, the monetary economy, and the growing professional armies. Montecuccoli's activity as head of the Austrian army and his writings, particularly the *War against the Turks in Hungary* composed at the same period, strikingly reflect these trends.

Among the developments resulting were both increasing supply problems and more elaborate logistic arrangements which, together with the cumbersome combination of pike and musket, and the effectiveness of field and permanent fortifications, made a strategic and tactical decision difficult to reach. Montecuccoli's campaign against Turenne won him his reputation as a master of manœuvre warfare. In his writings, he elaborated on supply considerations and warned against risking the country's army and fortune in battle, unless conditions were promising.[38]

[35] *Vom Kriege mit den Türken*, Bk. I, ch. I, Aphorism 1, *AS* ii. 206.

[36] *Von der Kriegskunst*, *AS* ii. 31.

[37] Ibid., and *Vom Kriege mit den Türken*, *AS* ii. 206. Also compare these with the works of Basta, Wallhausen, and Savorgnino cited above.

[38] *Abhandlung*, II. 3; *Vom Kriege mit den Türken*, I. 6, III. 6. Rothenberg, 'The Seventeenth Century' in Paret (ed.), *Makers of Modern Strategy*; G. Perjés, 'Army Provisioning, Logistics and Strategy in the Second Half of the 17th Century', *Acta Historicae Hungaricae*, XVI 1–2 (1970), 1–51; M. Roberts, 'The Military Revolution 1560–1660' in his *Essays in Swedish History* (London, 1967); G. Parker, 'The "Military Revolution 1560–1660"—a Myth?', *Journal of Modern History*, 48 (1976), 195–214; M. van Creveld, *Supplying War* (Cambridge, 1977), ch. 1.

The field of fortifications is of particular interest. Since the Italian wars in the late fifteenth and early sixteenth centuries and the rise of the new art of fortifications, developed to counter gunfire and based on the geometrical maximization of congruent fields of fire, the study of mathematics was widely regarded as fundamental for the study of war.[39] Lorini, the most distinguished Italian fortifications expert at the close of the sixteenth century, cited in Montecuccoli's source list, stated that fortifications were a science which, like medicine, was based on rules and mathematical considerations.[40] Indeed, corresponding to the *esprit géométrique* of the seventeenth century, the study of military geometry was a popular pastime at European courts.[41]

In *On the Art of War*, following an intellectual and educational pattern established by a whole series of military authors from the second half of the sixteenth century, Montecuccoli presented decimal arithmetic, the calculation of spaces, and trigonometry as necessary knowledge for the art of war, and devoted the first three chapters of the work to their systematic teaching.[42]

Montecuccoli's writings, probably regarded as a state secret and only circulated internally, were never published during his lifetime. However, when *On the War against the Turks in Hungary* or *Aphorisms*—as well as *On the Art of War*—appeared at the beginning of the eighteenth century, it was universally read. It was translated into all the major European languages (apart from English), and was published, within about a hundred years, in seven Italian, two Latin, two Spanish, six French, one Russian, and three German editions.[43]

The military thinkers of the Enlightenment were not that interested in Montecuccoli's military conceptions. On the contrary, with the

[39] For the emergence of geometrical fortifications, see C. Duffy, *Siege Warfare: The Fortress in the Early Modern World 1494–1660* (London, 1979). For the role of mathematics in the mind of military thinkers in the 16th cent. in connection with the new art of fortifications, see Hale, *Renaissance War Studies*, ch. 7, 'The argument of Some Title Pages of the Renaissance'.

[40] Jähns, *Geschichte der Kreigswissenschaften*, pp. 845–6.

[41] For the Austrian case, see Evans, *Habsburg Monarchy*, p. 334.

[42] For the literary pattern, see Hale in *New Cambridge Modern History*, iii. 178–9. For a comparison with Vauban's slightly later, epoch-making system, see Kaufmann, *Montecuccoli*, ch. 5.

[43] Rothenberg, 'The Seventeenth Century' in Paret (ed.), *Makers of Modern Strategy*, p. 60.

military developments of the eighteenth century these became quite outdated and, as we shall see, the military thinkers of the Enlightenment with their universal outlook found this somewhat disturbing. It was Montecuccoli's theoretical vision and conceptual framework that were widely admired and adopted. The men of the Enlightenment were, fortunately, not aware of the exact nature of Montecuccoli's scientific interests which probably would have horrified them. But his intellectual assumptions appeared familiar enough. It was not his particular form of science but the scientific outlook itself that counted. Montecuccoli worked out a sophisticated formulation of a new theoretical paradigm in the study of war, expressing a new, emerging world-view. Following the introduction of firearms, historical change was, to a limited degree, recognized; but it was overshadowed by the notion of universal rules and principles which was inspired by the sciences and reflected a new intellectual enterprise to subject all spheres of life to the domination of reason.

2
The Military Thinkers of the French Enlightenment

The Quest for a General Theory of War

Reflecting the Outlook of the Enlightenment

In the middle of the eighteenth century a sharp upsurge in the volume of military literature—reflecting an intense and unique intellectual activity—took place in Europe, spreading from France to the rest of the continent. Indeed, it may be instructive to start with some quantitative data. According to Pöhler's bibliographical survey of military works, more than seventy items were published, in an almost even distribution throughout the seventeenth century, in the 'art of war' category, which encompasses the more general and comprehensive theoretical works. A similar rate of publications was maintained in the first half of the eighteenth century with more than thirty items appearing in the years 1700–48. Then, between 1748–56, twenty-five items were published in a dramatic fourfold increase, and this rate was maintained in the period between the Seven Years War and the French Revolution (1756–89) with the publication of more than one hundred works. No substantial change in the number of publications occurred in the Napoleonic period or throughout the nineteenth century. The middle of the eighteenth century, therefore, marked a revolutionary growth in military publications.[1]

While this quantitative analysis shows the scope of the literary tide, it cannot reveal its origins and nature. In trying to explain these in

[1] J. Pöhler, *Bibliotheca historico-militaris* (Leipzig, 1887–97), iii. 583–610; the bibliographical items include new editions, which, like original publications, are indicative of the increasing literary activity. Such a quantitative analysis of an essentially qualitative matter is obviously crude, particularly as the distinction between works on the art of war, military history, tactics, and the various arms is fundamentally arbitrary, and the concepts themselves underwent considerable change in meaning. A similar trend is noticeable, however, in all the above-mentioned categories.

his monumental compendium of military literature, Jähns looked for answers from within the military sphere itself. The wars of Frederick the Great, he suggested, stimulated the awakening of military thinking and writing.[2]

This is hardly a satisfactory explanation. The upsurge in literary activity had taken place before the Seven Years War in which Frederick the Great earned his military reputation. Furthermore, the rate of military publications was hardly affected by any military event, be it the Thirty Years War, the wars of Louis XIV, or those of the Revolution and Napoleon. The flourishing of military literature from the middle of the eighteenth century and the ideas peculiar to it cannot be explained on military grounds.

Jähns's interpretation is merely indicative of the curious fate of one of the most influential schools of military thought, which dominated the eighteenth century, and whose legacy has since shaped the theoretical outlook on war.[3] The very existence of this school, not to mention its profound origins, collective ideas, and scope of influence, has hardly been recognized by modern scholars. Stemming from the all-encompassing ideas of the Enlightenment, which dominated all spheres of European thought and culture (including Frederick the Great's world-view), it closely followed the fortunes of the Enlightenment from its heyday to its eclipse.

This intellectual milieu can only be very roughly outlined here. On the accumulated strata of the doctrine of natural law, the neo-classical search for rules and principles in the arts, and Cartesianism, which together had dominated Louis XIV's France, stressing that reality was subject to universal order and to the mastery of reason, the gospel of Newtonian science was added. This gospel had conquered French culture by the 1730s largely owing to the support of Voltaire, its most influential propagator, and the far-reaching intellectual prospects that it appeared to have opened up affected

[2] Jähns, *Geschichte der Kriegswissenschaften*, p. 1451.

[3] Jähns's work with its invaluable, exhaustive account of primary sources is an astonishing example of 19th cent. German historical scholarship. However, following in Moltke's footsteps and in accordance with contemporary views, Jähns's conception of military *Wissenschaft* as organized, systematic knowledge based on clear concepts is modest (ibid., pp. v–vi). Coupled with the fact that Jähns is unaware of the general intellectual context of military thought (or indeed of any social or political context), this conception, imposed as it is on Jähns's subject-matter, is an unhistorical instrument, insensitive to the actual nature of the theoretical outlook of any particular period, especially the Enlightenment.

all sciences and arts. Most of the thinkers of the Enlightenment were not interested in physics as such and did not delve into the mathematical subtleties of science. Newtonian science was for them a symbol of the ability of the human mind to master reality, and they sought to extend its astonishing achievements to include the whole intellectual world. The scientific model was perceived by them as a general method for the foundation of all human knowledge and activity on an enduring basis of critical empiricism and reason.[4]

This common ideal overshadowed the many differences of opinion between the *philosophes*, who were mostly divided between deists and atheists, dualists and materialists, exponents of natural law and advocates of utility, believers in progress and primitivists, supporters of enlightened absolutism, aristocracy, and democracy—to mention the most notable differences. It was responsible for a remarkable degree of cohesion that encompassed all spheres of culture, and it was shared by a large educated community whose social élite mingled in the salons. This community embraced the ideal of universal knowledge, believed man could understand everything, and encouraged and approached with enthusiasm any new attempt to reveal the universal foundations of each discipline.[5]

Following in Locke's footsteps, Condillac developed associationist psychology in his *Essai sur l'origine des connaissances humaines* (1746), *Traité des systèmes* (1749), and *Traité des sensations* (1754). And Helvétius carried it in an hedonist direction and towards utilitarian ethics in his *De l'esprit* (1758). Society and politics were governed by principles that arose from the nature of things; Montesquieu's *De l'esprit des lois* (1748) expounded this in a manner that drew general admiration throughout Europe. Political economy was formed as a science in the 1750s with the activity of the physiocrats headed by Gournay, Quesnay, and Turgot. La Mettrie's *L'Homme machine* (1747) and Holbach's *Système de la nature* (1770) offered a materialist explanation of man and nature. Rousseau wrote his prize-winning essay *Discours sur les sciences et les arts* for the

[4] From the plethora of scholarly literature on the Enlightenment, recent and most extensive works are Peter Gay, *The Enlightenment: an Interpretation* (2 vols.; London, 1967–9), and Ira O. Wade, *The Structure and Form of the French Enlightenment* (2 vols.; Princeton, 1977); see also Paul Hazard, *European Thought in the Eighteenth Century* (London, 1954).

[5] For the ambivalent role of the salons and the social environment, see K. Martin, *French Liberal Thought in the Eighteenth Century* (London, 1954), 103–16.

Academy of Dijon in 1749–50, his *Du contrat social* appeared in 1762, and his *Émile* (1762) opened an era of educational theory. The *Encyclopédie* edited by Diderot and D'Alembert first came out in 1751, symbolizing the period; all spheres of human culture and all natural phenomena were to be subjugated to intellectual domination, and war was no exception.

Under the entry 'Guerre', Le Blond, a well-known fortifications expert, described the theory of war as being based on rules and principles derived from the experience of various generations. Military theory was founded by the classical authors and further developed by modern military thinkers, notably Montecuccoli and a series of more recent authors, most of whom were French.[6]

One of these was Antoine Manassès de Pas, Marquis de Feuquières (1648–1711), a lieutenant-general in the French army, whose widely read *Mémoires* (1711), translated into English and German, offered a set of military maxims in all branches of the conduct of war. These maxims were freely demonstrated by a perceptive analysis of cases taken from the wars of Louis XIV. Another author was the famous Jean Charles, Chevalier de Folard (1669–1752), whose works, particularly his *Histoire de Polybe* (1724–30) and related studies, calling for the revival of shock tactics, provoked interest throughout Europe.

As mentioned above, however, it was from the late 1740s that a new theoretical enterprise, unprecedented in scope and sense of vocations—emerged in a whole series of military works. At the height of the French Enlightenment, military thinkers incorporated the all-encompassing outlook of the period into the military field. War, they complained, was ruled by 'arbitrary traditions', 'blind prejudices', 'disorder and confusion'. All these had to be replaced by critical analyses and systematic schemes which the men of the period understood in definitive and universal terms, largely overriding circumstantial differences and historical change. The organization of armies and conduct of war would thus become an orderly discipline with clear theoretical tenets.

The ideal of Newtonian science excited the military thinkers of the Enlightenment and gave rise to an ever-present yearning to infuse the study of war with the maximum mathematical precision and certainty possible. However, the model that dominated their work

[6] Diderot and D'Alembert (edd.), *Encyclopédie*, vii (Paris, 1757), 823–6.

was less rigorous, and stemmed from the highly influential legacy of seventeenth-century neo-classicism in the arts.

The neo-classicists believed that they had found in Aristotle's *Poetics* a set of rules and principles for the construction and critique of artistic creation, among which the doctrine of the three unities and the rigid framework of genres were particularly influential. These rules and principles were embodied, according to the neo-classicists, in the works of the geniuses of the classical period on the one hand and of the age of Louis XIV on the other, which provided a universal standard of measurement to which all creative activity had to conform. From the beginning of the eighteenth century, this conceptual framework was being increasingly infused throughout Europe with a more liberating spirit, placing growing emphasis on the role of the creative imagination and the free operation of genius. This legacy dominated the arts in France until the late eighteenth century; and the arts dominated the Enlightenment, including its military facet.[7]

Indeed, the military thinkers of the Enlightenment maintained that the art of war was also susceptible to systematic formulation, based on rules and principles of universal validity which had been revealed in the campaigns of the great military leaders of history. At the same time, it escaped formalization in part, while the rules and principles themselves always required circumstantial application by the creative genius of the general.

De Saxe

Maurice de Saxe (1696–1750) was one of the many illegitimate sons of Frederick Augustus ('the Strong') of Saxony, later King of Poland. Early in his life, de Saxe became a soldier of fortune, and acquired his first military experience in the War of the Spanish Succession and, under Eugène of Savoy, in the war against the Turks. In 1720, he joined the French army, where he made a name for himself not only in the field but also in court and social circles. His continual efforts to procure an independent principality met with failure. But his victories at Fontenoy (1745), Raucoux (1746), and Laffeld (1747),

[7] For a general survey, see C. H. C. Wright, *French Classicism* (Cambridge Mass., 1920). Also see E. Cassirer's classical account in *The Philosophy of the Enlightenment* (Princeton, 1951), pt. VII; and R. Wellek, *A History of Modern Criticism* (London, 1955), vol. i.

as commander of the French invasion of Flanders during the latter part of the War of the Austrian Succession, gained him the rank of Marshal General of All the Armies of France and European renown as one of the greatest generals of the period.[8] His *Reveries on the Art of War* was widely circulated and aroused great interest. It is discussed here first despite a chronological ambiguity: though written in 1732, the book was only published posthumously in 1756 and was preceded by several other major expressions of the new quest for a general theory of war.

De Saxe's famous description of the state of military theory was transmitted through Jomini to the nineteenth century. It is an archetypal expression of the world-view of the Enlightenment: 'War is a science covered with shadows in whose obscurity one cannot move with an assured step. Routine and prejudice, the natural result of ignorance, are its foundation and support. All sciences have principles and rules; war has none.'[9]

Much as he regrets this situation, it may appear that de Saxe does not presume to change it. He seems openly to declare this at the beginning of his book: 'This work was not born from a desire to establish a new system [*système*] of the art of war; I composed it to amuse and instruct myself.'[10] Furthermore, the author himself appears to suggest that his book should not be taken too seriously. In a note to the readers he states: 'I wrote this book in thirteen nights. I was sick; thus it very probably shows the fever I had. This should supply my excuses for the irregularity of the arrangement, as well as for the inelegance of the style. I wrote militarily and to dissipate my boredom.'[11] What more can be said to belittle the significance

[8] De Saxe's life story and colourful love affairs, which did not lag behind his father's, have attracted some two dozen popular biographies, but a modern scholarly study is still missing. His campaigns up to 1746 were studied exhaustively by the Historical Branch of the French General Staff: see J. Colin, *Les Campagnes du Maréchal de Saxe*, (3 vols., Paris, 1901–6), followed by a fourth volume, *La Campagne du Maréchal de Saxe 1745–6*, by Henry Pichat (Paris, 1909). For his ideas on battle formation and deployment, see Quimby, *The Background of Napoleonic Warfare*, pp. 41–61.

[9] Since both English translations of the time (London, 1757, and Edinburgh, 1759) are unreliable, reference is made to the modern, albeit abridged translation of de Saxe's *Mes Rêveries* in T. Phillips (ed.), *Roots of Strategy* (Harrisburg, 1940); here, see p. 189. Wherever I deviate from this version, the French original (Amsterdam and Leipzig, 1757) is quoted in brackets.

[10] *Reveries*, p. 189.

[11] Ibid., p. 300.

of one's own work? The readers of the nineteenth and twentieth centuries who were removed from the values, literary forms and stylistic norms of the early modern period, could have received, with Jomini's courteous assistance, no other impression.

In fact, before the emergence of new attitudes during the Enlightenment, it was customary for authors to present themselves as casual scribblers and amateurs who wrote only incidently to 'amuse themselves' and 'ease their boredom'. The chivalrous ethos still dominated social values, and authors recoiled at the idea of being regarded as scholars. Thus, if we were to take their own account of themselves seriously, Montaigne, the brilliant essayist, scribbled with no attention to style, merely for himself and for the amusement of his family and friends; Montecuccoli wrote for himself, to clarify his own concepts; Feuquières wrote out of a fatherly devotion to his son's education; and de Saxe composed a work of some three hundred printed pages during thirteen nights of fever.[12] These accepted literary gestures should not be taken at face value.

It was convenient for Jomini to cite de Saxe's gloomy account of the state of military theory as evidence of the insignificance of all preceding theoretical work. Having done this, he now turned, for the same reason, to crush de Saxe's own work:

> The good Marshal Saxe, instead of piercing those obscurities of which he complained with so much justice, contented himself with writing systems for clothing soldiers in woolen blouses, for forming them upon four ranks, two of which to be armed with pikes; finally for proposing small field pieces which be named 'amusettes' and which truly merited that title.[13]

Ironically, in his treatment of his predecessors, Jomini only reiterated the characteristic attitude of almost all the military thinkers of the Enlightenment, including de Saxe. Since military theory was required to be definitive and universal, and since all past attempts were obviously not, they could only be perceived as failures. De Saxe himself dismissed all earlier theoretical work, though in a much subtler manner than Jomini. The great generals, he wrote, left no

[12] For this literary custom, its roots in the chivalrous system of values, and the case of Montaigne, see Peter Burke, *Montaigne* (Oxford, 1981), 3–4. For Montecuccoli see his *Ausgewählte Schriften*, i. 5; for Feuquières, see his *Memoirs Historical and Political* (London, 1736), vol. i, p. xxxii. For the change in the social ideal during the Enlightenment from the 'gentleman' to the 'bourgeois' and the 'philosophe', see P. Hazard, *The European Mind 1680–1715* (London, 1953), 319–34.

[13] Jomini, *Summary of the Art of War* (New York, 1854), 10.

instructive principles, and historians wrote on war from their imagination. Gustavus Adolphus established a military method in the organization of his army and was followed by many disciples. His contemporary Montecuccoli was the only one to examine the military profession in some detail. However, since the time of Gustavus Adolphus,

> there has been a gradual decline amongst us, which must be imputed to our having learned only his forms, without regard to principles . . . Thus there remains nothing but customs, the principles of which are unknown to us. Chevalier Folard had been the only one who has dared to pass the bounds of these prejudices,

but in the final analysis he too has been wrong.[14] Rather than shunning the theoretical challenge and indulging in trivialities, as Jomini implied, de Saxe was preparing the ground for his own military system.

In describing de Saxe's ideas, Jomini obviously selected the most marginal examples and those which appeared particularly peculiar in the 1830s. However, his scorn also throws light on de Saxe's conception of theory, which was similar to that of Montecuccoli and all the early military thinkers of the Enlightenment. A definitive military system was to encompass and determine all aspects of war down to the smallest details. Clausewitz described this as the first stage in the development of military theory:

> Formerly, the terms 'art of war' or 'science of war' were used to designate only the total body of knowledge and skill that was concerned with material factors. The design, production and use of weapons, the construction of fortifications and entrenchments, the internal organization of the army, and the mechanism of its movements constituted the substance of this knowledge and skill. All contributed to the establishment of an effective fighting force.[15]

This was precisely the nature of de Saxe's theoretical effort. 'The courage of the troops', he wrote, adapting the conceptual framework of neo-classicism, 'is so variable . . . that the true skill of a general consists in knowing how to guarantee it by his dispositions, his positions and those traits of genius that characterize great captains.' However, 'before enlarging too much upon the elevated [*elevées*] parts of war, it will be necessary to treat of the lesser, by which I mean

[14] *Reveries*, pp. 189–90.

[15] Clausewitz, *On War*, II, 2, p. 133.

the principles [*principes*] of the art'. As in architecture for example, the knowledge of the fundamental principles is a prerequisite to the operation of genius.[16] Accordingly, the first part of the *Reveries* deals with the 'details' of army organization, battle formation, armament, and so on, whereas the second part is concerned with the 'sublime parts' of war: all forms of warfare—in the open field, on mountains and rough terrain, during a seige, and against field fortifications—dominated by the general's genius.

De Saxe's work is a comprehensive treatise on war. He puts forward his ideal military model, his 'legion', and taking issue with the views and practices of his age, he advances many original ideas. However, rather than discussing his military doctrines, the aim of this book is to elucidate the intellectual premises that dominated his mind: he saw a need to subject military affairs to reasoned criticism and intellectual treatment, and the ensuing military doctrines were perceived as forming a definitive system.

The *Reveries* attracted much attention when it appeared in 1756. The author's fame contributed to this, and his ideas were widely discussed throughout Europe. The book was reprinted three times in 1757 alone, and again in 1761 and 1763. It was almost as widely circulated in German (1757, 1767) and English (1757, 1759, 1776) editions. A collection of miscellaneous military studies written by de Saxe shortly before his death, was published in 1762 under the title *Esprit des lois de la tactique*. The influence of Montesquieu is apparent both in the book's title and references. In one of the essays, 'Memoire militaire sur les Tartares et les Chinois', de Saxe responded to the emerging global view of the world and to the fashionable interest, stimulated by Voltaire, in the vast, remote, and ancient Chinese civilization.[17] In another essay, he discussed Marquis de Puységur's new book on military theory published in 1748.

Puységur

Jacques-François de Chastenet, Marquis de Puységur (1655–1743) began his long military career in the wars of Louis XIV during which

[16] *Reveries*, pp. 190–2.

[17] See esp. Basil Guy, 'The French Image of China before and after Voltaire', *Studies on Voltaire and the Eighteenth Century*, xxi (1963); J. H. Brumfitt, *Voltaire—Historian* (Oxford, 1958), esp. ch. II, pp. 76–84; Cassirer, *Philosophy of Enlightenment*, pp. 197–233.

he became Marshal Luxemburg's quartermaster-general (chief of staff), and finished it in the 1730s, in the War of the Polish Succession, as Marshal of France. His widely read *Art of War by Principles and Rules*, published posthumously in 1748 (reprinted 1749), was the first to propound the new ideal of a general theory of war, and was translated into German and Italian (1753).

Echoing Montecuccoli, Puységur wrote that war was the most important of sciences and arts, and yet lacked a systematic theoretical study, with people relying on tradition and personal experience. In his search for a theory of war, he reviewed the military works of antiquity, Turenne's memoirs, and Montecuccoli's writings, but found no satisfactory, comprehensive theory. In the *Art of War by Principles and Rules* he attempted to correct this state of affairs.[18]

The universal theory of war was to be derived from historical observation. Again using an argument of Montecuccoli's, Puységur dismissed the challenge of historical change. The introduction of firearms, he wrote, led some to believe that modern war was a new type of war for which the military theory of the ancients was no longer relevant. There could be no greater mistake. Despite all changes in armament, the science and art of war remained the same at all times. Expressing neo-classical conceptions, Puységur emphasized that the successes of all the great generals throughout history had been the result of adherence to the universal rules of war.[19]

The full scope of historical experience was therefore to be the source of military theory. Puységur's main interest lay in developing a system for the movements and deployment of armies, and for this the practices of antiquity were indeed still of considerable value. The works of Homer, Herodotus, Socrates, Xenophon, Thucydides, Polybius, Arrian, Plutarch, and Vegetius are cited in his *Art of War by Principles and Rules* together with those of Turenne and Montecuccoli.[20] Turenne's campaigns are compared to Caesar's.[21] Finally Puységur's own scheme is presented and then demonstrated

[18] Puységur, *Art de la guerre par principes et par régles* (Paris, 1748), *avant-propos*.

[19] Ibid., *avant-propos*.

[20] Ibid., pt. 1, chs. I and II.

[21] Ibid., pt. 2, chs. IV, VI, IX–XI.

in a fictitious campaign geographically set between the Seine and the Loire rivers.[22]

Alongside this fundamentally historical approach, Puységur also gave expression to a much more ambitious theoretical ideal. The perfection of the systematic siege warfare by Marshal Vauban, Louis the XIV's famous master of siegecraft, fascinated the military thinkers of the Enlightenment, for it exemplified the seemingly enormous potential of the *esprit géométrique* in the military field.

Vauban's highly renowned *De l'attaque et de la defense des places*, published in numerous editions, was the standard work for students of fortifications and siegecraft until the second half of the nineteenth century when these subjects were transformed by mechanization and developments in metallurgy and ballistics. Vauban perfected the geometrical system of fortifications and also developed a highly effective method of attacking fortresses. The besieging army with its sappers and guns approached the enemy fortifications through a system of earthworks, advancing in zigzag trenches and deploying in successive 'parallel' ones, thus being protected from the defender's fire. This was a systematic and uniform procedure that achieved an almost certain breakthrough with little bloodshed.[23]

As Clausewitz wrote in his account of the development of military theory: 'Siege warfare gave the first glimpse of the conduct of operations, of intellectual effort.'[24] Indeed, the military thinkers of the Enlightenment regarded it as an ideal to be expanded. Once conceived, the methods of fortifications and siegecraft provided a clear and exact—almost fully geometrical—guide for action, requiring only mechanical application. And if siege warfare was subject to a priori and precise reasoning, why could not the same be achieved in all branches of war?

This was the reasoning propounded by Puységur. Field warfare had to be made as scientific as siegecraft had been by Vauban. For

[22] For Puységur's ideas on marching and deployment, see Quimby, *The Background of Napoleonic Warfare*, pp. 16–25.

[23] Vauban, *Traité de l'attaque et de la defense des places* (2 vols., La Hare, 1737). For Vauban's life and military career, see Paul Lazard, *Vauban* (Paris, 1934) and Reginald Blomfield, *Sebastin Le Prestre de Vauban 1633–1707* (London, 1938). A good, concise account is Henry Guerlac's 'Vauban: The Impact of Science on War' in Earle (ed.), *Makers of Modern Strategy*. Also see C. Duffy, *Siege Warfare, 1494–1660*; id., *Fire and Stone, the Science of Fortress Warfare 1660–1860* (London, 1975); and id., *The Fortress in the Age of Vauban and Frederick the Great 1660–1789* (London, 1985).

[24] Clausewitz, *On War*, II, 2, p. 133.

this, emphasis had to be put on the study of geometry and geography and on their application to the art of war.[25] Armies operated in space, and while geography offered concrete knowledge of this space, geometry was to provide a precise instrument for analysing and regulating the movements of armies within it. This ideal attracted all the military thinkers of the Enlightenment, but was not pursued rigorously until Bülow.

Turpin de Crisse

Count Turpin de Crisse, a hussar officer and later a lieutenant-general, contributed extensively to the growth of military literature from the late 1740s. His many works included commentaries on Caesar (1769, 1785, and 1787), Montecuccoli (1769 and 1770), and Vegetius (1775, 1779, and 1783). His comprehensive *Essai sur l'art de la guerre* (1754 and 1757)—like the works of all the major military thinkers of the French Enlightenment—was well known throughout Europe and translated into German (1756 and 1785), English (1761), and Russian (1758).

In his theoretical outlook, Turpin was somewhat less radical than most of his contemporaries. Rather than blaming the rule of tradition and prejudice alone for the lack of systematic military theory, he pointed out the problems inherent in the subject-matter itself. In war, rules and principles were difficult to determine and hard to apply:

> Of most other sciences the principles are fixed . . . Philosophy, mathematics, architecture and many others are all founded upon invariable combinations. Every man, even of a narrow understanding, may remember rules [and] apply them properly . . . but the study of war is of another kind . . . nothing but a mind enlightened by a diligent study can make a due application of rules to circumstances.

Both genius and study are required.[26]

Though less precise and determinant than in other sciences and arts, the rules and principles of war were still absolutely and universally valid. 'The principles of war among all nations and in all times have been the same, but the little experience of the early ages of the world

[25] Puységur, *Art de la guerre*, p. 2.

[26] L. de Turpin de Crisse, *The Art of the War* (London, 1761), vol. i, pp. i–ii.

would not permit those principles to unfold themselves.'[27] The fundamental universalism of the military thinkers of the Enlightenment allowed no change in the essentials of the art of war or of military theory.

In his *Essai sur l'art de la guerre*, Turpin discusses extensively all branches of war very much after the manner of Montecuccoli. One of the last chapters of the treatise, entitled 'A Principle on which the Plan of Campaign May Be Established', is however of particular interest, being one of the earliest attempts to systematize the conduct of operations. Puységur posed the challenge of expanding the achievement of systematic siegecraft to field warfare. Turpin proposes a direct application of Vauban's celebrated technique. Indeed, he writes,

> Why could there not be some general method established which, being accommodated to the circumstances of time and place, would render the event of the operations more certain and their success less dubious? Art is now brought to that perfection, and there is almost a certainty of carrying a place when the siege of it is properly formed . . . it seems probable that the principles which serve for the conducting of a siege, may become rules for forming the plan . . . of campaign.[28]

A general must choose his objective and advance towards it until he meets with resistance. Then he should build a system of fortifications and depots across the front—the equivalent of the 'first parallel'. Having established that, he may resume his advance, zigzagging and forcing the enemy to withdraw. He should then establish a second system of fortifications and depots—the 'second parallel'. From there, he may again zigzag forward to a 'third parallel' which would already bring his objective within reach.[29] 'The success of campaign . . . based upon this maxim', wrote Turpin, 'seems to be almost certain.'[30] Furthermore, 'a general who proposes succeeding by such a method will find prudence more necessary than bravery.'[31]

Maizeroy

The poor performance of the French army in the Seven Years War stimulated intense intellectual activity within the French military

[27] Ibid. 183.
[28] Ibid. 99, 106.
[29] Ibid. 99–103.
[30] Ibid. 103.
[31] Ibid. 99.

up until the Revolution. The deep sense of inferiority in military organization and doctrine generated a willingness to carry out extensive experiments and reforms. These made the French army the most progressive in Europe, and forged many of the military instruments that were to be employed by the armies of the Revolution and Napoleon.[32]

The Seven Years War had established the Prussian army as the best in Europe. The generalship of Frederick the Great was universally admired, and its role in bringing about the Prussian successes was obvious. But genius was regarded as an intangible quality which could hardly be studied, and it was therefore the organization and doctrines of the Prussian army that attracted all the attention. It was believed that the Prussians had won their brilliant victories by perfecting almost mechanically the firing and manœuvring potential of the linear formation operating in close order. Consequently it was universally assumed that what the French army needed was a battle formation that would equal and even surpass the Prussian model. In the 1760s and particularly the 1770s, the efforts of the military thinkers of the French Enlightenment were thus totally concentrated on developing such a formation.

All agreed on one point: the French could not and ought not to compete with the automatic, almost inhuman perfection of the Prussian drill and battle order. From Folard to du Picq, Foch, and Grandmaison, French military thinkers held to the opinion that the French people were too volatile and 'had too much imagination' to be subjected to the iron discipline of the 'phlegmatic' Prussians, or to equal their perseverance. On the other hand, French enthusiasm, initiative, aggressiveness, and quarrelsome nature allowed for freer and more flexible doctrines.

In the 1720s, Folard revived the idea of the deep formation and shock tactics as a reaction against the triumph of linear formation and firepower. De Saxe too advocated more reliance on the *arme blanche*. And in the 1750s Folard's ideas were propagated by his disciple Mesnil-Durand in *Projet d'un ordre françois en tactique* (1755). Although these ideas were received with interest, they became

[32] The pioneering and admirable study of this development is J. L. A. Colin, *L'Infanterie au XVIIIe siècle* (Paris, 1907), see esp. ch. II, pp. 73–134. Also see Quimby, *The Background of Napoleonic Warfare*; Albert Latreille, *L'Armée et la nation à la fin de l'ancien régime* (Paris, 1914); and Émile G. Léonard, *L'armée et ses problèmes au XVIIIe siècle* (Paris, 1958), esp. chs. X and XII.

the focus of attention only after the Seven Years War when they sparked the great doctrinal controversy over the *ordre profond* and the *ordre mince*.

The French Enlightenment swarmed with definitive systems intended to regulate this or that sphere of human life, and, characteristically, all the participants in the intense military controversy believed that it was to produce a system of a definitive and absolute nature. As Clausewitz wrote in his outline of the development of military theory: 'tactics attempted to convert the structure of its component parts into a general system.'[33]

Paul Gideon Joly de Maizeroy (1719–1780) was a lieutenant-colonel in the French army when the first two volumes of his *Cours de tactique, théoretique, pratique et historique* appeared in 1766. These were followed by two complementary volumes (1767 and 1773) and by the *Theorie de la guerre* (1777). The *Cours* was reprinted twice (1776 and 1785) and translated into German (1767 and 1773) and English (1781). A well-known student of classifical warfare, Maizeroy became a member of the French Royal Academy of Inscriptions and *belles-lettres*.

Unlike many of his contemporaries, Maizeroy did not declare himself the founder of military science, which he already regarded as an established fact. It was true, he wrote, that the art of war in France had 'followed a blind and lazy routine',[34] but fortunately, 'in an enlightened and learned age in which so many men's eyes are employed in discovering the numerous abuses which prevail in every department of science and art, that of war has had its observers like the rest'.[35] Folard had been the first to work out and set down a military system,[36] developed by Mesnil-Durand. And Puységur, Turpin, and de Saxe had propounded their own systems. Maizeroy too had one to offer.

Historical study was the basis of military theory. Together with Guichard, Maizeroy was the most important expert of his time on the art of war in antiquity. He wrote several specialized works on

[33] Clausewitz, *On War*, II, 2, p. 133.

[34] P. G. Joly de Maizeroy, *A System of Tactics* (London, 1781); i, 357. This is a trans. of the first two vols. of the *Cours de tactique, théoretique, pratique et historique* (Paris, 1785).

[35] Ibid. ii. 179.

[36] Ibid. i. 357.

that subject, and devoted the first part of the *Cours* to a scholarly study of Greek and Roman warfare, which he compared with, and brought to bear on, modern warfare. Maizeroy also published the first French translation of *Tactica* (1770), the military treatise written in the ninth century by the Byzantine Emperor Leo on the basis of Emperor Maurice's sixth-century *Strategica*. Finally, reflecting the broadening of the historical and geographical scope beyond the boundaries of Europe, Maizeroy also examined, in the third part of the *Cours*, the warfare of the Turks and Asians.

Throughout history, Maizeroy believed, war conveyed clear lessons, provided it was seriously studied. 'The theory of the Greeks was fixed, certain and uniform, because it was treasured up in methodical treatises.'[37] No change could affect the universal fundamentals of the art of war:

> Though the invention of powder and of new arms have occasioned various changes in the mechanism of war, we are not to believe that it has had any great influence on the fundamental part of that science, nor on the great manœuvres. The art of directing the great operations is still the same.[38]

Adapting neo-classical conceptions, de Saxe had distinguished between the fundamental part of war, governed by rules and principles, and the sublime part; and Turpin had stated that war, while based on rules and principles, required a great deal of creative application. Maizeroy elaborated on this intellectual framework. One part of war, the

> merely mechanical, which comprehends the composing and ordering of troops, with the manner of encamping, marching, manœuvring and fighting . . . may be deduced from principles and taught by rules; the other [is] quite sublime and residing solely in the head of the general, as depending on time, place and other circumstances, which are eternally varying, so as never to be twice the same in all respects.[39]

The construction of armies and their combat doctrines constitutes the sphere of 'tactics'. The meaning of this concept in the eighteenth century has often been unclear to later readers. Deriving from the Greeks the concept was rarely used until the eighteenth century. With the revival of interest in classical warfare, stimulated by Folard,

[37] Maizeroy, *Cours de tactique*, i. 361.
[38] Ibid., p. viii. [39] Ibid. ii. 353.

the concept became popular with the military thinkers of the Enlightenment, and has since been a central military technical term, though changing slightly in meaning. The military thinkers of the Enlightenment used it in the Greek original sense to mean a system of army organization and battle formation. However, in the 1760s and particularly in the 1770s, as they became engrossed in this field and regarded it as the core of military theory, they also used 'tactics' as a general term for the art of war as a whole. Furthermore, since they tended to look upon the conduct of armies on the battlefield predominantly as a product of their battle formation and related doctrines, 'tactics' also implied the conduct of battle itself.[40] Only at the end of the century, with Bülow, did the emphasis in the concept change and assume its current meaning as the art of conducting battle.

The search for the perfect system of tactics is therefore the principal theme in Maizeroy's writings. In the controversy over the column and the line, Maizeroy held a moderate position in favour of the *ordre profond*. He regarded his position to be a direct conclusion from the universal nature of military theory; any doctrine had to be based on the experience of the Greeks and Romans as well as on contemporary conditions. Those who maintained that the invention of firearms rendered deep formation obsolete, implied that war was a craft rather than a science, because they disregarded a universal principle—the importance of depth for cohesion and morale.[41]

Furthermore, the principles of tactics were not only universally valid but also based upon the rigorous and precise rationale of mathematics. Explicitly referring to the Pythagorean philosophy that numbers underlay all phenomena, Maizeroy maintained that military formation had to be based on the correct choice of the universal numbers that insured flexible internal division and manœuvre. Odd numbers, for instance, prevented the subdivision into two equal parts.[42]

The conduct of operations was the second branch of the art of war. Maizeroy gave this branch a new technical term, 'strategy', whose

[40] Compare Le Blond's article 'Tactique', in the *Encyclopédie*, xv (Paris, 1765), 823–6.

[41] Maizeroy, *Cours de tactique*, iv. 13.

[42] Ibid. iv. 21–4.

origins in modern military theory also seem to have been lost. Maizeroy, who translated the Byzantine military classics into French, was the one who introduced the concept that derived from the Greek word for general and was used by Emperor Maurice as the title for his military treatise *Strategicon*. Maizeroy employed it for the first time in 1777 in his later work *Theorie de la guerre*. The concept was slow in penetrating French military jargon and was still almost unknown in Britain at the beginning of the nineteenth century. In Germany, however, where Maizeroy was widely read, and where a German translation of Leo's was published in 1781, the term was rapidly accepted and already incorporated into the military literature of the 1780s. Bülow divided the conduct of operations between strategy and tactics in the sense which is known today, and through his works and German military literature this usage was accepted throughout Europe during the nineteenth century.[43]

To what extent then can strategy be reduced to precise and universal rules and principles? In 1777 Maizeroy expounded upon his positions of 1766, which had already been implied by de Saxe and Turpin. Strategy belongs

> to the most sublime faculty of mind, to reason. Tactics is easily reduced to firm rules because it is entirely geometrical like fortifications. Strategy appears to be much less susceptible to this, since it is dependent upon innumerable circumstances—physical, political, and moral—which are never the same and which are entirely the domain of genius. Nevertheless, there exist some general rules which can be determined safely and regarded as immutable.[44]

These rules of strategy (also called the 'military dialectic' by Maizeroy) are:

> not to do what one's enemy appears to desire; to identify the enemy's principal objective in order not to be misled by his diversions; always to be ready to disrupt his initiatives without being dominated by them; to maintain a general freedom of movement for foreseen plans and for those to which circumstances may give rise; to engage one's adversary in his daring enterprises and critical moments without compromising one's own position; to be always in control of the engagement by choosing the right time and place.

[43] Bülow, *The Spirit of the Modern System of War* (London, 1806), 86–7. The translator's note (p. 34) indicates that the term strategy was still virtually unknown in Britain.

[44] Maizeroy, *Théorie de la guerre* (Nancy, 1777), pp. lxxxv–lxxxvi.

To these are also added: 'not to deviate from one's main objective . . . [and] to secure one's communications'.[45] Maizeroy demonstrates these principles through an analysis of several campaigns of great French generals.

Maizeroy's principles of strategy—remarkably similar to the twentieth-century abstract notion of the principles of war—were quite an isolated theoretical structure. At the end of the eighteenth century when interest was to focus on discovering the rationale of operations, the search was to be for much more concrete and meaningful principles. And Maizeroy's own period of writing in the 1770s was totally dominated by the quest for a definitive system of tactics.

Guibert

The intensive doctrinal fermentation in the French army following the Seven Years War reached its climax in the 1770s with the great controversy over the *ordre profond* and *ordre mince*. Guibert's *Essai général de tactique* appeared at the beginning of that decade and won the admiration of the salons. Mesnil-Durand launched a fierce counter-attack in his *Fragments de tactique* (1774), once again propounding the superiority of the column and shock action. An extensive testing of his system was conducted in the camp of Vaussieux (1778) under the supervision of Marshal Broglie, the foremost soldier in France. And one year later Guibert published his *Defense du système de guerre moderne* (1779). The *philosophes* attention was attracted to military theory as a result of this controversy, and especially owing to the work and personality of Guibert.

Jacques Antoine Hippolyte Comte de Guibert (1743–90) embodied the remarkable integration of military theory with its intellectual environment. He was a child of the Enlightenment through and through, embedded in its cultural achievements, sharing its characteristic ideas, and stimulated by its particular code of values and standards of excellence. At the age of twenty-six he had already composed a tragedy in verse, *Le Connétable de Bourbon* (1769), but he achieved his meteoric renown in his own professional field a year later when he published his military treatise, the *Essai général de tactique*.

[45] Ibid. 304–5.

Guibert grew up under the strong influence of his father's military career. The elder Guibert was Marshal Broglie's right-hand man in the Seven Years War, and assisted him in carrying out his famous military reforms, including the introduction of a proto-divisional system. After the war, he was responsible for developing combat formations and drills in the War Office. The young Guibert joined the army as a small boy. He participated in the campaigns of the Seven Years War, and later in the war in Corsica (1768), and rose to the rank of colonel by the age of twenty-six. Receiving his military education personally from his father, and serving with his staff in the latter part of the Seven Years War, Guibert became deeply involved in his father's interests and preoccupations. Closely familiar with official French military thinking and planning, he appeared on the scene of military theory.[46]

The *Essai général de tactique* was obviously first and foremost a contemporary military work, and it was Guibert's brilliant propositions in the military sphere that aroused great interest in professional circles and made his book one of the most influential military treatises of the eighteenth century. But it was Guibert's belief—characteristic of the period—that his work offered a definitive system of tactics, finally creating a science of war, and it was the comprehensive expression that he gave to the ideas of the Enlightenment, that made his book a success with the *philosophes* and the talk of the salons. Guibert wrote the *Essai* with a pronounced and conscious intention to create an immortal masterpiece; this is apparent in every line of his work. His intellectual environment clearly determined not only the nature and strength of this desire but also the attitudes and themes required for its realization. The ambitious and enthusiastic young man appeared to have incorporated into his military treatise as many ideas of the Enlightenment as possible and touched upon most of its major concerns.

[46] No comprehensive biography of Guibert has yet been written. See F. E. Toulongeon's introd. to Guibert's *Journal d'un voyage en Allemagne 1773*, in Guibert's *Œuvres* (Paris, 1803), published by his widow; Flavien D'Aldéguier, *Discours sur la vie et les écrits de Guibert* (Paris, 1855); R. R. Palmer, 'Frederick the Great, Guibert, Bülow', in Earle (ed.) *Makers of Modern Strategy*, repr. in Paret (ed.), *Makers of Modern Strategy*; and Lucien Poirier, *Les Voix de la stratégie—Guibert* (Paris, 1977). For the most interesting testimonies on Guibert's period of glory, see below.

The introduction of the *Essai* opens in defence of the *philosophes* against the accusation that they threaten the foundations of society and particularly that they undermine patriotism.[47] A 'Preliminary Discourse' beginning with 'A Review of Modern Politics' was mainly responsible for the success of the *Essai* in the salons. Firstly, modern political and social institutions are compared with those of antiquity, and Guibert takes sides with the ancients in the famous controversy that spanned the seventeenth and eighteenth centuries in France.[48] In a well-known passage—later to be regarded as prophetic—he asserts that the best political and military constitution and an enormous potential of power are embodied in the vital institutions of the republic of the masses, drawn in the image of the ideal, simple, and vigorous republics of antiquity.[49]

Unfortunately, modern Europe appears to be too corrupt and degenerate to rise to this model. Indeed, 'what . . . do the politics of Europe present to a philosophic mind disposed to contemplate them? Tyrannical ignorance or weak administrations.'[50] Like most of the *philosophes*, Guibert therefore places his hopes on enlightened absolutism:[51] 'some moral and philosophical truths which gradually filter through error, will by degrees unfold themselves; at last one day or other reach a sovereign . . . and render posterity more happy.'[52] Frederick the Great, the friend and hope of the *philosophes*,[53] is Guibert's natural hero. Prussia's political and military institutions as well as the personality of its king are highly praised, both in the *Essai* and in later works.[54]

[47] J. A. H. Guibert, *A General Essay on Tactics* (London, 1781), vol. i, p. vi.

[48] For a general account of the controversy between the 'Ancients' and the 'Moderns', see J. B. Bury, *The Idea of Progress* (New York, 1932), ch. 4; and O. A. Oldridge, 'Ancients and Moderns in the Eighteenth Century', in P. Wiener (ed.), *Dictionary of the History of Ideas, Studies of Selective Pivotal Ideas* (New York, 1968), i. 76–87.

[49] Guibert, *Essay*, p. viii.

[50] Ibid., p. iv.

[51] For the *philosophes'* attitude toward enlightened despotism that changed from hope to disappointment, see esp. Martin, *French Liberal Thought*, pp. 132–42; Hazard, *European Thought*, pp. 325–34; and Gay, *The Enlightenment*, i. 483–97.

[52] *Essay*, p. 4.

[53] See n. 51 above, and for the particular case of Voltaire see P. Gay, *Voltaire's Politics* (Princeton, 1959), pp. 144–70.

[54] See esp. Guibert's *Observations on the Military Establishment and Discipline of the King of Prussia* (Berlin, 1777; English trans., London, 1780); and *Éloge du Roi de Prusse* (London, 1787).

Montesquieu revealed to the men of the Enlightenment a new depth of connection between all the elements of the socio-political fabric. Guibert responds to the challenge:

Politics is naturally divided into two parts, *interior* and *exterior* politics. The first is the basis for the second. All which belongs to the happiness and strength of a people springs from their sources, laws, manners, customs, prejudice, national spirit, justice, police, population, agriculture, trade, revenues of the nation, expenses of government, duties [and] application of their produce.[55]

A comprehensive scientific study of the politico-military sphere must, therefore, analyse all these factors in depth. Guibert explains that he has not yet carried this out in the *Essai général de tactique*, and this is why he calls it 'general'. But he does intend to take upon himself the writing of this extensive work, which is to be called 'A Complete Course of Tactics'. He even presents the full outline of this work (which he was never to write). It is to open with an analysis of the political constitutions of all the European countries (thirty-four in all). The domestic politics of each of these countries is to be examined in view of all the above-mentioned factors, while their foreign policy is to be studied in relation to each other. Only then will all the elements of military science itself be discussed: 'Elementary Tactics' deals with the various arms, and 'Great Tactics' deals with marching, combat deployment, and encamping.[56]

From the political background Guibert proceeds to discuss war itself. First, 'A Review of the Art of War Since the Beginning of the World' extends the new universal view of history to the military sphere.[57] Then the main problem is presented: the state of the science of war. The ambitious young man acknowledges no predecessors. As usual, all competitors are brushed aside with a thoroughness only to be equalled by Jomini. Guibert alleges that the great generals of history left no principles. Works in military history are inaccurate and, in particular, do not provide guidance; they do not point out 'causes and effects'. There are also some didactic works such as those of Caesar, Rohan, Montecuccoli, Marshal de Saxe, and the King of Prussia, but they too are deficient; they are not detailed and explicit enough. Finally, there are modern writers, but

[55] *Essay*, p. xxi.
[56] Ibid., pp. lxxviii ff.
[57] Ibid., p. xvi.

who can be recommended? Folard? Puységur, who is full of errors? Guichard, who dealt mostly with antiquity? De Saxe? As to the writers who are still alive, Guibert's attitude is somewhat different; here one must be more generous or, perhaps, cautious. He declares that he has, of course, no intention of offending Turpin, Mesnil-Durand, or Maizeroy; he has learnt a great deal from them.[58]

Something fundamental is very wrong in the science of war:

> Almost all sciences have certain or fixed elements, which succeeding ages have only extended and developed, but the tactics, till now wavering and uncertain, confined to time, arms, customs, all the physical and moral qualities of a people, have of course been obliged to vary without end and for a space of a century to leave behind them nothing else but principles disavowed and unpracticed, which have ever been cancelled and destroyed by the following age.[59]

The tension between historical change and circumstantial differences on the one hand and the dominating universal view inspired by the scientific ideal on the other, was inherent in the minds of the military thinkers of the Enlightenment.

Military science, Guibert asserts, must adopt the methods that brought success in other sciences. The works of Newton, Leibnitz, and D'Alembert are the models to be followed.[60] Guibert is not satisfied with anything less than the top of mathematical science. Incorrect methodology, he maintains, rather than the nature of the subject-matter itself, has been responsible for the failure of military theory:

> Let us suppose that the first mathematical truths are taught to a people inhabiting the two extremes of the globe . . . they must evidently in time arrive at the same result of principles. But has there been in the tactics any clear truth demonstrated? Are the fundamental principles of this science established? Has one age ever agreed on this point with its preceding one? But why was there no such work, which could have laid a firm foundation for its principles? It is for this reason that the military have for a long time been ignorant how to analyse the subject . . . and unacquainted with the method of explaining and arranging their ideas.[61]

[58] Ibid., pp. xlvi–xlviii.
[59] Ibid. 1.
[60] Ibid., p. xxvii.
[61] Ibid. 2–3.

Guibert's system of tactics is to solve the confusion and lay down definitive principles of universal validity. Then,

> the tactics . . . would constitute a science at every period of time, in every place, and among every species of arms; that is to say, if ever by some revolution among the nature of our arms which it is not possible to foresee, the order of depth should be again adapted, there would be no necessity in putting the same in practice to change either manœuvre or constitution.[62]

Guibert's system of tactics will thus settle all theoretical differences and establish a clear guide for action. The radical and ambitious young man finds the great works of the Enlightenment somewhat deficient in this respect; they leave the reader with no definite solution to the question of how to proceed. Montesquieu's masterpiece is one example; Helvétius's and the *Grande Encyclopédie* are another.[63] Science should be advanced to encompass everything:

> It would be very interesting to see military science improve . . . in this manner . . . I have already remarked how the same revolution be made in politics. This maxim would likewise take place in almost all the sciences, provided their theory was divested of all those errors . . of false methods . . . Then the encyclopedia of human understanding, now becoming the repository of truth, would assume her reign and affirm herself amidst the various alterations of ages.[64]

It is hardly surprising that Guibert's book was received warmly by the *philosophes* and in the salons.

The intellectual circles in which Guibert's work was highly acclaimed were neither particularly interested in, nor knowledgeable about, military affairs. But its military worth was in any case recognized widely, and it was its general intellectual connotations that impressed the laymen. Guibert's contemporaries were accustomed to the publication of masterpieces that laid the foundation of one sphere or another of human life and thought. The *Essai général de tactique* was accepted as one of these works. 'M. de Guibert', wrote the celebrated literary critic Sainte-Beuve more than a century later,

> was a young colonel for whom society . . . roused itself to a pitch of enthusiasm. He . . . published an 'Essay on Tactics' preceded by a survey

[62] *Essay*, p. 99. [63] Ibid., p. lxviii. [64] Ibid., p. lviii.

of the state of political and military science in Europe . . . He competed at the Academy on subjects of patriotic eulogy; he had tragedies in his desk on national subjects. 'He aimed at nothing less', said La Harpe, 'than replacing Turenne, Corneille and Bossuet'. [He was] a man whom every one, beginning with Voltaire, considered at his dawn as vowed to glory and grandeur . . . you will not find a writer of his day who does not use the word [genius] in relation to him.[65]

Guibert's success in the Parisian intellectual circles was indeed spectacular. Mlle de Lespinasse, who hosted one of the most important salons in the capital and was a close friend of D'Alembert, fell in love with Guibert. The hundreds of letters she wrote to him between 1773 and 1776 when she died of a 'broken heart' after he married another woman, vividly evoke the Parisian intellectual environment and Guibert's success, aspirations, and connections with the *philosophes*.[66] Another great mistress of the salons and intimate of Guibert, Mme de Staël, wrote *Éloge de Monsieur de Guibert* after his death, in which she attempted to explain his failure to fulfil the hopes placed upon him in his youth.[67] The poem *La Tactique*, written by Voltaire after the publication of the *Essai*, is another vivid testimony to the social success of the work and to the impression left by the personality of its author, particularly in view of Voltaire's ambivalent attitudes to the subject of the work itself.

The patriarch of the Enlightenment was throughout his life a bitter enemy of war. Frederick the Great's unscrupulous use of military means was one of the major factors that cooled relations between Voltaire and the philosopher-king. In a series of works of which *Candide* was only the most famous, Voltaire never tired of denouncing war, blaming it on the cynical ambition of rulers and the folly of peoples.[68] He may also have made the theory of war a target for his irony by saying that 'the art of war is like that of

[65] Introd. to Mlle J. de Lespinasse, *Letters* (London, 1902), 8–9.

[66] Ibid. The letters were published by Guibert's widow in 1809; a complete edn., including some of Guibert's own letters, was published by a descendant of Guibert in 1906, and translated into English in 1929.

[67] Madame la Baronne de Staël, *Œuvres complètes* (Paris, 1821), xvii, 275–317, ed. by her son.

[68] For Voltaire's views on war, see ch. XI of Léonard's excellent *L'Armée au XVIIIe siècle*, which includes a great deal of material on the *philosophes'* attitude to war. Also see Gay, *Voltaire's Politics*, pp. 160–1; and Martin, *French Liberal Thought*, pp. 265–7.

medicine, murderous and conjectural'.[69] But when the *Essai général de tactique* was published, Voltaire did not doubt that this was indeed a general theory of war, and a new achievement of the Age of Reason.

The poem *La Tactique* (1774) opens with a bookseller showing Voltaire the new work:

> Tactics, says I, I do declare, till now
> Not half their worth and value did I know.
> 'The name' he answered 'came from Greece to France;'
>
> . . .
>
> I therefore shut the door and read it through,
> Intent to gain by heart, with instant labour,
> The Art, my friends—to kill my neighbour.
>
> . . .
>
> Strangely surprised at this so boasted Art,
> Back I returned to CAILLE [the bookseller] with
>   wounded heart,
> And throwing him his book, in warmth, I said,
> 'Go, thou by Satan for his uses made,
> The Tactics give to the Chevalier de Tot
>
> . . .
>
> But first to FREDERIC bestow
>   your skill,
> And be assured he knows its meaning well;
>
> . . .
>
> A greater murderer than the great Eugene,
> Or great Gustavus
>
> . . .
>
> Thus I express'd myself—while, listening nigh,
> A youth had mark'd me with a curious eye;
> His uniform two epaulets did grace,
> Which his profession, and his rank express;
> His mien was steady, tranquil, and serene,
> His talents, not his courage, there were seen;

[69] I could not locate this quotation, cited without reference in J. Fuller, *The Foundations of the Science of War* (London, 1925), 19.

In short, it was the Author of the book,
He thus accosted me, with modest look:
'I can perceive' says he 'you disapprove;
you are an old Philosopher, and love
Mankind entire—This Art is not humane,
But needful to the earth, I say't with pain,
Where many an ABEL has a brother CAIN'.

In the poem Guibert goes on to pose the question how history would have looked had the civilized nations from Rome to France not defended themselves against the barbarians, and what the fate of culture and of the fatherland would have been.[70]

I made not a reply—the truth I saw,
And felt the force of reason's sovereign law.
I look'd on War the first of human Arts;
On him . . .
Who made the science, in his numbers, swell
Fit to command, in what he knew so well,

Yet, in my breast, I own, there rose a sigh,
I wish'd the Art, from want of use, might die;
That equity, on earth might bring to bear
Th'ideal peace, o'the Abbé de la Saint-Pierre.[71]

The *Essai* went through four editions in five years (1770, 1772, 1773, 1775), and was translated into German (1774) and English (1781). In a journey to Germany in 1773, Guibert met Frederick the Great and Joseph II.[72] In 1785 he was elected member of the

[70] For Voltaire's general attitude that regarded wars in defence of one's country and of civilization as a necessary evil, see Gay, *Voltaire's Politics*, pp. 160–1.

[71] 'Tactics', in Voltaire, *Works*, trans. T. Smollett (London, 1779–81), Misc., i. 126–30.

[72] Frederick casually referred to the *Essai* and to Guibert, his admirer, in his correspondence with D'Alembert and Voltaire. In a letter to Voltaire in which he complained about the intellectual poverty of the generation, Frederick humorously described the books sent for him by his literary agents from Paris: 'a book has been published on the art of shaving dedicated to Louis XV . . . essays on tactics are written by young officers who know not how to spell Vegetius' (The King to Voltaire, 16 Jan. 1773, in Frederick, *Posthumous Works*, trans. T. Holcroft (London, 1789), viii. 249–50). This remark is, however, perhaps too incidental to be indicative of the king's attitude; it may be an example of his famous cynicism. For a neutral and even more incidental reference to Guibert, see: The King to D'Alembert, 17 Sept. 1772, ibid. xi. 318.

French Royal Society of Sciences. He continued his military career and military, political, historical, and dramatic writings until his death during the early stages of the Revolution.

'It would be very easy at this date, but not very just, to make a caricature of M. de Guibert.'[73] When Sainte-Beuve wrote these words at the end of the nineteenth century, Guibert's reputation was at its lowest ebb, and he was mostly remembered as a short-lived celebrity and the lover of Mlle de Lespinasse. However, Sainte-Beuve's words were soon to become much more meaningful than he himself intended. Indeed, Guibert's theoretical aspirations may now appear boundless and his burning ambition amusing. But such a view would ignore the intellectual context in which he operated—the world-view, vision, and ideals of the men of the French Enlightenment. Moreover, at the beginning of the twentieth century, Colin's classical studies of the origins of French warfare under the Revolution and Napolean revealed the full influence of Guibert's military ideas which can only be touched upon here.

Perhaps Guibert did not create the one definitive general military system, but he wrote a superb doctrinal work which greatly influenced the development of future warfare. He propounded revolutionary ideas: mobility, rapidity, and boldness in the conduct of operations; the solving of logistical problems by a massive reliance on the countryside; movement in independent formations, similar to the proto-divisional system introduced by Marshal Broglie; and flexible manœuvring in open columns before deploying into the firing-line, instead of the highly complex and rigid manœuvring of the linear formation that had been employed and perfected by the Prussians. These ideas flowed into the melting-pot of the dynamic French military thinking of the last years of the *ancien régime*, and moulded the doctrines of the French army on the eve of the Revolution. Guibert's ideas were practically the basis of the official Ordinance of 1791 with which the armies of the Revolution went to war, and the *Essai* played a major role in the military education of Napoleon.[74]

[73] Introd. to Mlle de Lespinasse, *Letters*, pp. 8–9.

[74] In addition to the works cited in n. 32 above, see J. Colin, *La Tactique et la discipline dans les armées de la Révolution* (Paris, 1902); and id., *L'Éducation militaire de Napoléon* (Paris, 1901).

After the wars of the Revolution and during the Napoleonic period, when military thinkers began to analyse new experiences and challenges, they did so—despite the overwhelming revolution in military reality and perceptions—in the light of the dominant theoretical ideal that had been spread throughout Europe by the major military thinkers of the French Enlightenment. In Germany this ideal was carried forward by the military thinkers of the *Aufklärung*, who were initially only a provincial group heavily influenced by the ideas from France, the centre of culture, but who later applied these ideas in new and revolutionary directions.

3

The Military Thinkers of the German Aufklärung

The development of military thought in Germany during the last third of the eighteenth century fits remarkably well into the general pattern that characterized the career of the Enlightenment in Europe. Initially, the military school of the German *Aufklärung* was overshadowed by its senior French counterpart. It emerged as a significant movement only in the 1770s, a generation after the theoretical developments in France. And although it originated independently, from a cultural environment similar to the French, and bore distinctive intellectual characteristics, the German school was influenced decisively by the major military thinkers of the French Enlightenment. However, after a period of growth during the 1780s, the intellectual enterprise of the military *Aufklärer*s took off towards the end of the century in novel, if not radical directions, winning attention throughout Europe. Finally, it contributed dialectically to the emergence of new, formidable theoretical trends that, in the context of a general reaction against the ideas of the Enlightenment, rejected the intellectual premises that had guided the military thinkers of that period, but nevertheless continued, though in a redefined form, to follow their dominant ideal—the search for a general theory of war.

I THE MILITARY *AUFKLÄRERS*

The full impact of the Enlightenment on the military field—which still deserves to be studied—transcends the scope and aims of this work. Because of the relatively provincial character of the military school of the *Aufklärung* during its formative period, confined as it was to the linguistic boundaries of Germany, an exhaustive survey of the many military writers who operated in the 1770s and 1780s will not be attempted here either.[1] This chapter only outlines some of the major expressions of the influence of the Enlightenment on the military field, and focuses on the most notable exponents of its ideas and on their distinctive message. Scharnhorst's life story and intellectual notions—an excellent case-study which will be treated in the second part of this book for reasons of later historical developments—provide an additional insight into the period.

The subtle differences between the character and intellectual trends of the German Enlightenment and its French counterpart also found expression in the military sphere. Whereas the building of systems was the driving force behind the military thinkers of the French Enlightenment, the early military thinkers of the *Aufklärung* were motivated by a more humanistic vision with a strong educational emphasis. In France the creation of a military science was at the centre of the intellectual inquiry, and the quest was for a general and definitive formula. In Germany the scientific ideal was at first less rigorous—perceived as a systematic broadening of military knowledge—and most of the attention was concentrated on disseminating that knowledge throughout the wider circles of the

[1] A gold-mine of information on this subject is contained in the third vol. of Jähns's *Geschichte der Kriegswissenschaften*. The limitations of Jähns's treatment of this material, particularly his total unawareness of the intellectual background of the developments that he describes, have, however, already been mentioned. A rich variety of primary sources is also incorporated in Reinhard Höhn's *Revolution, Heer, Kriegsbild* (Darmstadt, 1944), which deals with the intellectual transformation involved in the transition from the warfare of the *ancien régime* to the wars of Revolution, and discusses extensively some of the trends described in this chapter. This is a very valuable work despite some difficult problems (see P. Paret, *Yorck and the Era of the Prussian Reform* (Princeton, 1966), 283–4), particularly Höhn's selective and sometimes inaccurate use of the sources, often harnessed to support a stereotyped argument. Various themes are also discussed in W. O. Shanahan, 'Enlightenment and War: Austro-Prussian Military Practice 1760–1790', in G. Rothenberg, B. Király, and P. F. Sugar (edd.), *War and Society in East Central Europe*, ii (New York, 1982), 82–111.

officer corps. This is mainly characteristic of the ideas and activities of the first military *Aufklärers* in the 1770s, but also has a bearing on the theoretical outlook of Frederick the Great, the pre-eminent representative of the Enlightenment in Germany.

The image of Frederick the Great in relation to the Enlightenment was somewhat ambivalent and underwent considerable transformation not only in the public's view but also in the military sphere. On the one hand, he was the hero of the military thinkers of the Enlightenment. While in philosophy, the sciences, and the arts, the philosopher-king sought the company of Voltaire, D'Alembert, La Mettrie, and Maupertuis, in the military field it was he who was the most important authority in Europe, admired as the foremost genius of the period and as the creator of a highly renowned military system. His military works and regulations and the institutions that he developed and established for the instruction of his officers also reflected the ideas of the Enlightenment.

Yet, on the other hand, Frederick's attitude to the rank and file, whom he regarded as fodder for his war machine and upon whom he imposed machine-like conduct and brutal discipline, appalled Voltaire, and aroused disapproval among some of the military even during the king's reign.[2] After the wars of the Revolution and the appearance of France's national conscripts who were motivated by patriotic and ideological sentiments, this limited disapproval turned into a deluge of criticism against the Frederickian military system. And following the defeat of 1806, with the activities of the reformists in the Prussian army, Frederick's system became synonymous with all that was outdated and inadequate in the *ancien régime*.

A similar development occurred in the field of military education. The king's activities in this field fell short of the programmes envisaged by the military *Aufklärers* who, from the 1770s, were calling for an improvement in officers' education, and for its extension into the ranks. As a result of these changing perspectives, Frederick has been portrayed as a military reactionary more than as an exponent of the world-view of the Enlightenment.

[2] For a very early example see the anonymous 'Versuch von der Kriegeszucht' in *Krieges Bibliothek*, I (Breslau, 1755); this periodical was edited by Georg Dietrich v. Gröben who was also perhaps the author of the essay; cited by Paret, *Yorck*, p. 18.

Frederick's military writings were composed when the intellectual enterprise of the military thinkers of the French Enlightenment was still in its infancy. His *Principes généraux de la guerre* or *Military Instruction for his Generals* had already been written in 1746 after the War of the Austrian Succession, while the *Elements de castramétrie et de tactique*, his most comprehensive military work, was composed in 1770 before Guibert—his admirer and the most forceful exponent of the theoretical ideas from France—created a sensation with his *Essai*. Unlike the histories, verse, philosophical and political essays, and aesthetical and dramatical critiques which the 'philosopher of *sans-souci*', as the king called himself, wrote and published for his pleasure, Frederick's military works were written with a clear practical aim and safeguarded like any other state paper. Only his *Instructions for his Generals* which fell into the hands of the Austrians in 1760, was published immediately in German and French (1761), English, Spanish, and Swedish (1762).[3]

Still, Frederick's military writings were just as clear an expression of the ideas of the Enlightenment as the works of the French military thinkers or his own unofficial writings. They reflected the belief that the art of war, like all arts, required a professional education and considerable knowledge, and could be treated theoretically on the basis of rules and principles that relied on historical evidence, could be used as a partial substitute for direct experience, and should be applied to particular cases through critical judgement.

At the opening of his *Eléments de castramétrie et de tactique* the king wrote:

> Those who are persuaded that valour alone suffices for the general officer, deceive themselves greatly; it is an essential quality, no doubt, but it must be matched with much other knowledge . . . [The general] must use judgment in everything and how can he do this if he lacks knowledge?[4]

The officer must have 'perfect knowledge of tactics or the art of manœuvre, of attacks, defences, retreats, marches, crossing rivers, convoys [and] forages . . . He must posses full knowledge of the

[3] Frederick's military writings were published in vols. 28–30 of the *Œuvres de Frédéric le grand* (Berlin, 1856). The German trans., vol. vi of *Die Werke Friedrichs des Grossen* (Berlin, 1913) also includes the *Militärische Testament* not authorized for publication in the original edn. There is also an earlier German trans. of the *Militärische Schriften* (Berlin, 1882).

[4] *Œuvres*, 29. 4.

county . . . field fortifications . . terrain . . . [and] defence and attack of fortresses.[5]

Frederick listed these branches of knowledge in his introduction to de Quincy's *Histoire militaire du règne de Louis XIV*, which was only one of the military classics that he ordered to be translated and distributed among his officers. Other works included Feuquières's *Memoires* and extracts from Folard's *Histoire de Polybe.* In his introduction to the latter, Frederick gave expression to the characteristic sentiment of the period: 'The art of war, which certainly deserves to be studied and investigated as much as any of the other arts, still lacks classic works.' He alleged that Caesar's works taught very little, and nothing of value was left from the late Roman empire. The art of war had been reborn only in the modern period with Maurice of Orange.[6]

> Every art has its rules and maxims; they must be studied. Theory facilitates practice. The lifetime of one man is not sufficiently long to enable him to acquire perfect knowledge and experience; theory helps to supplement it; it provides a youth with early experience and makes him skilful also through the mistakes of others. In the profession of war the rules of the art are never transgressed without punishment from the enemy.[7]

The neo-classical conceptual framework is apparent: 'it is only after repeated examination of what one has done that the artists succeed in understanding principles . . . Such research is the product of the applied mind.'[8] The king himself repeatedly synthesized his political and military experience for the benefit of his successors. 'I have seen enough', he wrote in his *Military Testament*, 'to offer general rules which are of special application in Prussia.'[9]

Applied thinking is always required because experience never repeats itself in exactly the same manner. In the introduction to his *History of the Seven Years War*, Frederick wrote:

> It is not probable that any similar chain of causes should, in a short time, produce the same circumstances as those under which we were . . . generals are never placed in exactly similar situations . . . past facts are good to store

[5] *Œuvres*, 29. 58, *avant-propos* (1771).

[6] Ibid. 28. 112, *avant-propos* (1753).

[7] Ibid. 29. 58–9.

[8] Ibid. 28. 169, 'Reflexions sur la tactique et sur quelques parties de la guerre' (1758).

[9] *Werke*, vi. 246.

in the imagination and the memory; they furnish a repository of ideas whence a supply of materials may be obtained, but which ought to be purified by passing through the strainer of judgment.[10]

In addition to the works of instruction that he circulated among his officers, and the seminars that he conducted for them, Frederick expanded and reorganized the cadet corps and established the *Académie militaire*, an officer academy with a broad general programme of studies which he sent to D'Alembert for his assessment.[11] These educational enterprises coincided with the appearance of military schools throughout Europe—one of the major indications of the influence of the Enlightenment.

The proliferation of military schools is undoubtedly connected to the rise of the absolutist state and the growth of central administration, but its roots go much deeper. The dominance of the absolutist state and the expansion of the standing professional armies predated the appearance of military schools. What was at work here was the emergence of the new idea that the military profession could be studied theoretically, and therefore required academic instruction; furthermore, that a broad general education was also essential for developing the officer's personality.

The most notable expression of this idea was the establishment of academies for officers alongside earlier and expanded cadet corps. The *École royal militaire* in France and the *Militär-Akademie* in the Austrian Empire were founded in 1752. The *Académie militaire* also known as the *Académie des nobles*, was created by Frederick in 1765. In Württemberg the military academy which had been formed in the early 1770s was incorporated into the new *Karls hohe Schule*. A *Militär-Akademie* was also founded in Bavaria (1789). Finally, Britain followed suit with the Royal Military College (1799), later at Sandhurst (1812), and the United States founded West Point in 1802. Because of Germany's political fragmentation and the flourishing of the German universities in the eighteenth century,[12]

[10] *The History of the Seven Years War*, *Posthumous Works*, vol. ii, pp. ix, xi, xii.

[11] The King to D'Alembert, 24 Mar. 1765, ibid. xi. 23.

[12] For this relative vitality and proliferation (nearly 50 in number), compared with the decline of the universities in France (22) and England (2), see T. C. W. Blanning, *Reform and Revolution in Mainz 1743–1803* (Cambridge, 1974), 11–12. For an extensive study and critical view see Charles E. McClelland, *State, Society and University in Germany, 1700–1914* (Cambridge, 1980), pt. I.

the academic idea was particularly in evidence in Germany. Even small German states such as Hessen-Hanau (1771) and Münster (1767) established military academies. Perhaps the most interesting case was in the tiny principality of Schaumberg-Lippe where Count Wilhelm, a general, man of the Enlightenment, and military *Aufklärer*, founded a military academy in 1766, where the cadet Scharnhorst received his earliest military education.[13]

Alongside the various types of officer schools there also appeared professional schools for the various arms, particularly for officers and NCOs of artillery and military engineering. These professions were universally regarded as scientific and the need for systematic theoretical training for them was recognized far more than for any other form of military education. Schools of engineering and artillery were founded or centralized in France (1749 and 1756 respectively), Austria (1717, reorganized 1755; 1786), Britain—the Woolwich Academy—(1741), Prussia (1788 and 1791), Saxony (1743 and 1766), Bavaria (1752 and 1786), and Hanover (1782 and 1786). The affiliation of West Point to the United States' corps of engineers is well known. Tempelhoff commanded the artillery school in Prussia, and the young Scharnhorst became an instructor at the newly formed artillery school in Hanover.[14]

These new military schools, however, still trained only part of the officer corps in Germany, and many officers continued to enter service without any regular training. Thus, perhaps the most interesting development was the spontaneous mushrooming of regimental military schools throughout Germany from the late 1770s. They were a product of both the proliferation of the state academies and the military literature of the Enlightenment discussed below. By and large, they were founded on the independent initiative of regiment commanders, exponents of the military *Aufklärung*, many of whom were graduates of the official academies. Usually the instructors in the school who taught the junior officers and often the NCOs were the senior officers of the regiment. With his transfer

[13] This outline is based upon K. von Poten's *Geschichte des Militär-Erziehungs und Bildungswesens in den Landen deutscher Zunge*, vols. 10, 11, 15, 17, and 18 of C. Kehrbach (ed.), *Monumenta Germaniae Pedagogica* (Berlin, 1889–97); and Jähns, *Kriegswissenschaften*, pp. 2447–92. For an 18th-cent. account by one of the military *Aufklärers* see F. Miller, *Reine Taktik* (Stuttgart 1787–8), i. 79–100. Also see Le Blonde, 'Études militaires' in the *Encyclopédie*, vi (1756), 94–6.

[14] See n. 13 above.

to the Hanoverian service in 1778, Scharnhorst, the graduate of Count Wilhelm's military academy, became an instructor in the military school established by the regiment commander, Colonel von Estorff, a notable representative of the growing circle of military *Aufklärers*. During his period in the Neuruppin garrison, the young Clausewitz may have participated in the activities of the school created in 1799 for the NCOs of his regiment by Colonel von Tschammer, a strong believer in the importance of education, and administrated by Major von Sydow, himself one of the first graduates of the *Académie militaire*.[15]

There was therefore a close interplay between the proliferation of military academies and schools and the emergence and expansion of a community of officers, who were advocates of the idea of military science and education, and who maintained intensive intellectual intercourse through extensive military literature that flourished in Germany from the late 1770s, cutting across its internal boundaries. The pioneering works of Ferdinand Friedrich von Nicolai and Friedrich Willhelm von Zanthier which appeared towards the middle of the 1770s, became the intellectual platform of this community.

The emphasis on education—typical of the Enlightenment belief in the ability to transform man and society and in the value of knowledge—was particularly popular during the German *Aufklärung*, which centred on the universities and which was highly influenced by pietism, more humanist and less political than its French counterpart.[16] While disciples of Rousseau such as Basedow, Salzmann, Rochow, Richter, and Pestalozzi were writing about, and experimenting with, the naturalistic approach to the education of children, military writers were stressing the necessity of military study and developing programmes for the education (*Bildung*) of officers.

Ferdinand Friedrich von Nicolai (1730–1806; not to be confused with the more famous exponent of the Enlightenment) was a colonel and staff officer in the Württembergian army. In 1769, on the request of Duke Carl Eugen and in collaboration with the University of Tübingen, he developed an educational programme for the planned

[15] Paret, *Clausewitz*, pp. 52–3; for the emergence in Prussia of regimental schools for the general education of the soldier's children, see ibid. 46–51.

[16] The best concise treatment of the differences between the French Enlightenment and the German *Aufklärung* is perhaps Blanning's 'The German Problem in the Eighteenth Century', pt. I of his *Reform and Revolution*.

military academy.[17] His ideas and experience were later offered to the public in the widely read *An Attempt at an Outline for the Education of Officers* (1773 and 1775).

The prevailing view, Nicolai wrote at the opening of his book, regards war as an art bestowed by nature on men of special talent. If war was a science of principles with mathematical foundations requiring theoretical study, the arguement goes, how could we explain Condé's victory at Rocroi when he was only twenty-two years, and had no previous military education? This view, Nicolai believed, characteristic of the Enlightenment, was the product of prejudice, ignorance, and the rule of tradition.[18] The ancients had a clear and fixed military science. And though war has changed owing to the invention of gunpowder, Maurice of Nassau and Gustavus Adolphus succeeded in their studies, and Montecuccoli re-established the science of war which was further developed by later thinkers.[19]

These opening theoretical statements are followed by the main subject of the work—a comprehensive programme for the education of officers. Military education alone is not sufficient; it must be preceded by a broad general curriculum to educate the man within the officer. Firstly, basic education is to be provided, including religion, languages, art, and the classics, followed by the advanced studies that include pure and applied sciences as well as history, geography, statistics, logic, ethics, and the laws of nature, nations, and war. For each of these disciplines Nicolai proposes a detailed programme of study and extensive bibliography. Finally, the military sciences themselves may be studied including: (*a*) equipment, organization, and armament; (*b*) military architecture; and (*c*) tactics, the science of warfare. Among these, the art of fortifications has achieved the highest scientific and mathematical status; Nicolai himself wrote a book on the subject, *Essai d'architecture militaire* (1755). The recommended reading material for the military sciences encompasses ancient and modern military classics as well as the works of all the major military thinkers of the French Enlightenment.

At about the same period, the characteristic themes of the new scientific-educational vision were also being propounded in Friedrich

[17] Poten, *Militär-Erziehungs*, 18. 316; for the academy see text above.

[18] F. Nicolai, *Versuch eines Grundrisses zur Bildung des Offiziers* (Ulm, 1775), 1–2.

[19] Ibid. 10–11.

Wilhelm von Zanthier's *An Attempt to Study the Art of War* (1775). All sciences, Zanthier writes, echoing Montecuccoli, de Saxe, and Guibert, have their textbooks and scholars, but the science of war has none. There are some works, but a general system is lacking. If war is to be studied as a science rather than a craft, theory above all must bring order into this labyrinth by clearly defining its various branches. Accordingly, Zanthier suggests the themes to be studied—battle deployment, marching, operational planning, camping, crossing rivers, and establishing winter quarters—and points out the principles of each.[20]

The works of Nicolai and Zanthier pioneered a wave of similar works that, from the late 1770s, stressed the scientific nature of war, the need for systematizing its study, and the necessity of military education.[21] Perhaps the military *Aufklärer*s did not comprise a majority in the officer corps, but they were obviously the officers who expressed their ideas in writing. It might therefore be more interesting to note a rare literary reaction against their increasingly influential views, which articulated traditional feelings and attitudes. The message of Leopold Schönberg von Brenckenhoff's little book, *Paradoxa, gröstentheils militärischen Inhalts*, which was published in several editions (1780, 1783, 1798), may be briefly summarized. Firstly, Brenckenhoff asserted that war was a craft to be experienced, hence all the difficulties in treating it theoretically. Second, and perhaps even more thought-provoking, was his claim that rather than advancing the military profession, education was probably harmful to military virtues. 'Philosophy clarifies our mind and makes us better human beings, but worse soldiers.'[22] Given a choice between an

[20] F. Zanthier, *Versuch über die Kunst den Krieg zu studiren* (n.p., 1775), 3–4 ff.; also see his later *Versuch über die Märsche der Armeen, die Läger, Schlachten und der Operations Plan* (Dresden, 1778).

[21] For a survey of works and authors, see: Jähns, *Kriegswissenschaften* pp. 2439–45; Höhn, *Revolution, Heer, Kriegsbild*, pp. 90–103. One characteristic work is Col. J. von Scholten's *Was muss ein Offizier wissen?* (Dessau and Leipzig, 1782), an address to the Society of the Friends of the Sciences and Good Taste. But the most interesting example is perhaps F. Nockhern von Schorn's *Versuch über ein allgemeines System aller militairischen Kentnisse* (Nuremberg, 1785; French edn., 1783). Nockhern von Schorn, a colonel in the Dutch army and an amazingly pretentious man, was clearly influenced by Kant's philosophy and fame, and ventured to generate a philosophical revolution in the study of war. His theoretical gospel, definitions, and educational programme deeply impressed Jähns, as they corresponded to his own conception of military science; Jähns, *Kriegswissenschaften*, pp. 1775–9.

[22] L. Brenckenhoff, *Paradoxa, grösentheils militärischen Inhalts* (n.p., 1783), 11.

army of savages and an army of educated troops whose officers are experts in the sciences and philosophy, Brenckenhoff stated that he would prefer the former.[23]

Brenckenhoff's doubts were not shared by the growing circle of officers who advocated the idea of military science and education, and whose activities throughout Germany became quite distinct by the late 1770s. The principal mouthpiece of this circle was the literary organ which it pioneered and popularized—the military periodical. As noted by Jähns, if France was the leader in all spheres of military thought, the emergence and flourishing of the military periodical was almost unique to Germany.[24] These publications reflected the general proliferation of periodicals in the Germany of the *Aufklärung*, which was related to a sharp increase in book production and the rapid expansion of the reading public.[25] Against the background of total political fragmentation, the military periodicals functioned as a means of intellectual communication for officers of similar interests beyond the restricted frameworks of the particular armies in which they served.

Gröben's *Kriegsbibliothek*, appearing from 1755, was the first military periodical in Europe. Twenty-three issues were published with a short intermission and several changes of title until 1784. Johann Georg Estor's *Sammlung militärischer Abhandlungen* appeared in Frankfurt am Main in 1763. Military periodicals did not truly flourish, however, until the late 1770s. The *Kriegerisches Wochenblatt* was published in Berlin in 1778. *Der Soldat* appeared in Hamburg in 1779/80, and the *Militärische Taschenbuch* was published in Leipzig in 1780. Andreas Böhnn, professor of mathematics in the University of Giessen and a disciple of the famous philosopher Christian Wolff, published a professional military periodical for officers of engineering and artillery, *Magazin für Ingenieurs und Artilleristen* (twelve issues in 1777–89). The young Scharnhorst, one of the most notable military *Aufklärer*s, edited some of the principal periodicals of the time. The four issues of his *Militär*

23 Brenckenhoff, *Paradoxa*, pp. 11–12.

24 Jähns, *Kriegswissenschaften*, p. 1812.

25 The number of both published books and writers practically doubled itself every decade in the last third of the 1770s; Albert Ward, *Book Production, Fiction and the German Reading Public, 1740–1800* (Oxford, 1974), esp. pp. 64, 167–8; Jonathan B. Knudsen, *Justus Möser and the German Enlightenment* (Cambridge, 1986), 145.

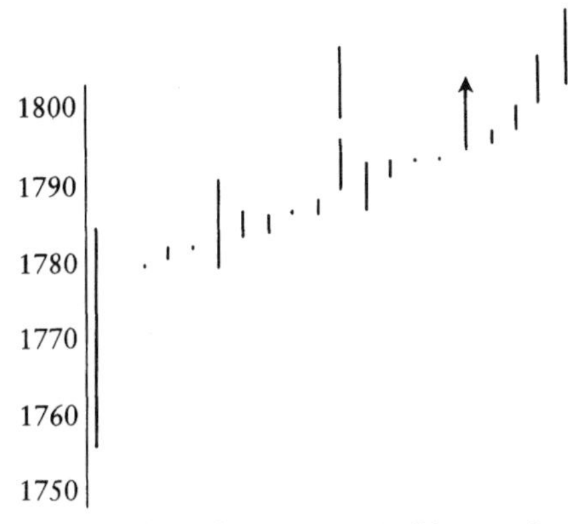

Fig. 1 Military Periodicals in the German *Aufklärung*: Spread and Duration of Publication

Bibliothek appeared in 1782–4, followed in 1785 by four issues of the *Bibliothek für Offiziere*. His *Neues Militärisches Journal* appeared between 1788 and 1793, resuming publication after the wars of the Revolution (1797–1805).

During the 1780s and 1790s there appeared Walter's *Bellona* (Dresden, 1781–5); de Stamford's and Massenbach's *Militärische Monatsschfrift* (Berlin, 1785–7); Oesfeld's *Genealogischer militärischer Kalender* (Berlin, 1784–90); Schleicher's *Neue militärische Bibliothek* (Marburg, 1789–90); *Neue militärische Briefe und Aufsätze* (Breslau, 1790); Schwerin's *Soldatenwesen* (Berlin, 1789); *Kleine militärische Bibliothek* (Breslau, 1790); *Archiv für Aufklärung über das Soldatenwesen* (Berlin 1792–3); *Der Österreichische Militär-Almanach* (Vienna, 1791); Küster's *Offizier-Lesebuch* (Berlin, 1793–7); *Berliner militär Kalender* (Berlin 1797–1803); and Hoyer's *Neue militärische Magazin* (1798–1808) (see Figure 1).[26]

A comparison with France, where only three short-lived military journals were published during the same period, highlights the volume of military literature in Germany.[27] The subscription lists, for example, which appear at the opening of Scharnhorst's periodicals, and which include hundreds of officers from all

[26] Jähns, *Kriegswissenschaften*, pp. 1812–23.
[27] Ibid. 1823.

over Germany, also attest to the scope and character of the reading public.[28]

During the 1780s the military periodicals reflected the relatively tranquil times. The professional and technical articles corresponded to the moderate conception of military science, aiming at the expansion of knowledge, and the discussion mainly reflected the great doctrinal controversies in France. This picture was transformed radically in the 1790s. The wars of the Revolution and the appearance of the French Revolutionary armies threw military thought in Germany into fierce debate between the guardians of the warfare of the *ancien régime* and the Frederickian military system on the one hand, and the advocates of the new military practices of the Revolution on the other.

Even prior to these developments, the controversial publication in Germany of the works of Lloyd, another disciple of the military thinkers of the French Enlightenment, stimulated renewed interest in the campaigns of the Seven Years War. The new focus on the conduct of operations led German military thinking in new, and even radical, theoretical directions. Until the 1790s, the military literature of the *Aufklärung* had been by and large confined to Germany's linguistic boundaries, and overshadowed by its French counterpart. Now the works of Lloyd, Tempelhoff, and Bülow attracted attention throughout Europe, and were translated into all the major languages of the continent.

[28] This subject still deserves to be studied; for Scharnhorst's periodicals, see below, ch. 6. I.

II LLOYD: HIS INTERNATIONAL CAREER, INTELLECTUAL SCOPE, AND THE CAMPAIGNS OF THE SEVEN YEARS WAR

The almost legendary life story of Henry Humphrey Evans Lloyd (*c.* 1718–83) has many facets. As a soldier of fortune, he served most of the political causes of Europe both on the battlefield and in clandestine operations. Deeply rooted in the Enlightenment and influenced by its great thinkers, he wrote extensively on his many interests. He adapted the theoretical teaching of the French military school to a German theme, and became the only British military thinker (if indeed he can be called one) until Fuller and Liddell Hart to influence the development of European military thought. Yet, he has received relatively little attention in his own country, and the full details of his life still remain unknown.[1]

Lloyd was born to a clergyman in a small village in north Wales and was educated at Jesus College, Oxford. Attracted to the military profession but unable financially to purchase a commission, he entered the clergy. In 1744, he went to France where, in a Jesuit college, he privately tutored officers in geography and field engineering, subjects on which he was already considered an expert. A year later, he took the first opportunity to leave the church and join the French army, in whose ranks he took part in the Battle of Fontenoy. His excellent drawings and ground survey of the battlefield attracted the attention of the army's chief engineer who awarded him a junior commission in the engineering corps. When the 'Young Pretender' prepared for his invasion of Scotland, Lloyd joined the expedition as a third engineer with a rank of captain. However, after the landing in Scotland, he was despatched to Wales carrying letters

[1] John Drummond, the son of a Scottish family that supported the Stewarts, and a friend of Lloyd in France between 1744 and 1756, wrote a short account of his life for the fifth edn. of Lloyd's *A Political and Military Rhapsody on the Invasion and Defence of Great Britain and Ireland* (London, 1798); see pp. ix–xii (this publication constitutes a later edn. of *A Rhapsody of the Present System of French Politics, of the Projected Invasion, and the means to Defeat It* (London, 1779)). Lloyd's son, the philologist and translator Hannibal Evans Lloyd, wrote a similar, though naturally somewhat biased, introd. to an even later edn. of this work (1842). The entry in *The Dictionary of National Biography (DNB)*, 1301–2 is mostly based on these two sources. Much new and exciting information and hitherto unknown works of Lloyd himself have been introduced by Franco Venturi in his excellent 'Le avventure del generale Henry Lloyd', *Rivista Storia Italiana (RSI)*, xci (1979), 369–433; for Lloyd's probable date of birth, see p. 369.

from the 'Pretender' to his supporters. This mission was the beginning of his international espionage career. Disguised as a clergyman, he left Wales and conducted a survey of England's southern sea-shores, preparing the groundwork for a French invasion. His activities aroused suspicion and he was arrested and transported to London. Fortunately for him, his participation in the expedition to Scotland was not discovered, and in 1747 he was released and returned to France.[2]

Promoted to the rank of major, Lloyd distinguished himself as an engineering officer at the siege of Bergen-op-Zoom. After the peace of 1748, he sought his fortune in the Prussian army, but fell back with the French in 1754 when new plans for the invasion of Britain were being drawn up, and again offered his services to the French Minister of War, Marshal Belle-Isle. He crossed the channel in 1756, and, this time disguised as a merchant, carried out his second extensive reconnaissance of British shores. However, as the plan to invade was abandoned, he sought military action elsewhere, joined the Austrian army, and was posted in Marshal Lacy's staff. Promoted to the rank of lieutenant-colonel, he participated in the first campaigns of the Seven Years War, and in 1760 commanded a reconnaissance force that followed the movements of the Prussian army.

During this period he met Pietro Verri (1728–97), the famous Milanese exponent of the Enlightenment, political economist, and man of letters, who served at that time (1759) as a captain in the Imperial army. The two men became friends, conducting long conversations and patrolling together. Verri was deeply impressed by Lloyd's intellectual breadth, varied interests, talents, and military expertise, and, according to Venturi, his works clearly reveal the influence of Lloyd.[3] In 1760, personal differences made Lloyd switch sides and join the service of the Duke of Brunswick. When the war in Germany ended, he attempted to join Count Wilhelm Schaumberg-Lippe who was defending Britain's ally Portugal against a French-supported Spanish invasion. He corresponded with the Count, sending him a political and military 'Memoir on the present state of Portugal'.[4]

[2] He probably travelled under an assumed name because no record of his imprisonment exists; *DNB* xi. 1301.

[3] Venturi, *RSI* xci. 374–5. Lloyd again met Verri in 1768–70 during his mission in Italy: ibid. 394–400.

[4] Ibid. 376–81. For Count Wilhelm, see pp. 157–8 below.

However, the hostilities in Portugal ended, and Lloyd resettled in England and embarked on an extensive literary career which was interrupted several times. In 1768 he went to Italy on a secret mission, this time serving the British government in an attempt to organize supplies for the defenders of Corsica against the French invasion.[5] In 1773–4 he accepted a Russian invitation to command a division against the Turks, and was promoted to the rank of major-general. In 1779, during the American War of Independence he composed his widely read *Rhapsody of the Present System of French Politics, of the Projected Invasion and the Means to Defeat It*, this time intending to point out ways to prevent a French invasion. Although the government paid his heirs not to publish further editions of this work, it was nevertheless published under a slightly different title during the invasion scare of 1794 and appeared in several later editions during subsequent invasion scares. According to his son, in 1782 Lloyd was intended to assume command in North America, but this claim seems doubtful.[6] He died in The Hague in 1783 and British agents are said to have conducted a search of his house and removed certain papers.[7]

The first volume of Lloyd's *The History of the Late War in Germany between the King of Prussia and the Empress of Germany and her Allies* appeared in London in 1766 with an extensive theoretical and programmatic introduction. His 'Reflections on the Principles of the Art of War' also known as 'Political and Military Memoirs' was published as a continuation of this volume in the second edition of the *History* (1781). The second volume of the *History*, compiled from Lloyd's papers, appeared posthumously in 1784. All these works were translated extensively in many different forms. The first volume appeared in at least three German and three French translations.[8] The *Memoirs* were brought out in no less than five

[5] According to the *DNB*, xi. 1302, there is no record of Lloyd's alleged governmental pension that was mentioned by Drummond (p. xiii of Lloyd's *Rhapsody on the Invasion and Defence of GB*), and it may have been Secret Service money.

[6] Hannibal E. Lloyd, introd. to Lloyd's *Political and Military Rhapsody* (1842 edn.), 9.

[7] *DNB* xi. 1302.

[8] Frankfurt and Leipzig 1777, Brunswick 1777 and 1779; London and Brussels 1784 and 1803, Lausanne 1784.

German and three French editions.[9] And Lloyd's complete work was translated into German by Tempelhoff (1783–94). This may still be an incomplete account.

Most elements of Lloyd's theoretical conception reflected the ideas propounded by the French military school, but since that school has fallen into oblivion, this fact has not been recognized by modern readers. These much rehashed themes will therefore be recounted here briefly, simply to show Lloyd's clear affinity to his predecessors, particularly to de Saxe. In his introduction to the *History* (1766), Lloyd writes that works on war, both historical and didactic, are unsatisfactory. The former are inaccurate and not elaborate enough, and the latter are too abstract. His own work combines the two forms.[10] Though very difficult to study, war, like all sciences and arts, is based upon fixed and invariable rules and principles. These comprise the mechanical part of the art and largely lend themselves to mathematical formulation. However, they require application to changing circumstances: this is the sublime part of the art which cannot be studied, and falls totally in the province of creative genius. As in poetry and rhetoric, principles are useless without divine fire.[11]

Lloyd's principles relate to the organization of armies. The first deals, for example, with the clothing of the troops, the second with shooting, the third with marching and deploying.[12] As mentioned, many of the principles of war are susceptible to mathematical formulation; Lloyd had distinguished himself as a military engineer from his youth. Fortifications are 'purely geometrical . . . and may therefore be learnt by anyone'.[13] This is also the case with artillery, which 'is nothing but geometry', and with the art of encamping.[14] Mathematical principles are also essential for calculating marches, which are based on considerations of time and space. They are even necessary for determining battle formations since 'the impulse that bodies, animate or inanimate, make on each other . . . is in proportion to mass and velocity'.[15]

[9] Frankfurt and Leipzig 1783, Münster 1783, Vienna 1785, Leipzig 1789 and 1802; London 1784, Basle 1798, Paris 1801.

[10] *History of the Late War between the King of Prussia and the Empress of Germany and her Allies* (London, 1781), vol. i introd., i–iv.

[11] Ibid., pp. vi–viii. [12] Ibid., pp. viii ff. [13] Ibid., pp. xxi–xxii.

[14] Ibid., pp. xxiii, xxi. [15] Ibid., pp. xx–xxi.

This far-reaching mechanistic position is not coincidental. Lloyd, who often described the army as a great machine, strongly adhered to the mechanistic and materialistic interpretation of the world. 'The modern philosophy,' he wrote, 'though for the most part founded on mathematical principles, has not in the course of a century been able to expel entirely the dreams and visions of Plato and Aristotle.'[16]

Lloyd's mechanistic outlook is fully revealed in his more extensive theoretical essay, 'Reflections on the Principles of the Art of War', or 'Memoirs' (1781). After the customary comparison between the ancients and the moderns, leading to the conclusion that shock-arms and the *ordre profond* should be incorporated more widely, Lloyd moves on to one of his major theoretical contributions to the military school of the Enlightenment. He is the first to develop his predecessors' notions regarding the moral qualities of the troops into a systematic study by applying the mechanistic-hedonistic psychology of the Enlightenment to the military field.

Thanks to Venturi's recent discoveries, we know that in the late 1760s after writing the first volume of his *History*, Lloyd wrote a substantial manuscript, 'Essai philosophique sur les gouvernements', which he probably intended to expand into a larger work 'on the different governments established among mankind'.[17] He apparently referred to this work when he told Piero Verri in 1768 that he intended to write a book which would be inspired by Helvétius and Montesquieu.[18] The influence of the former dominates chapters 2 and 3 of the 'Essai philosophique'—'Des sensations' and 'Des passions'—and it is again manifest in Lloyd's psychological discussion in the 'Memoirs', entitled 'The Philosophy of War'.

In the footsteps of Hobbes's *Leviathan* (1651), La Mettrie's *L'Homme machine* (1747), and Helvétius's *De l'esprit* (1758), Lloyd writes: 'Fear of, and an aversion to pain, and the desire for pleasure, are the spring and cause of all actions, both in man and other species of animals . . . Pain and pleasure arise from interior and mechanical causes.'[19] He discusses at length the emotions motivating generals and troops,

[16] Ibid. 12.

[17] The manuscript is deposited at the Fitzwilliam Museum in Cambridge; Venturi, *RSI* xci. 383. For the intended expansion see Lloyd's anonymous *An Essay on the English Constitution* (London, 1770), preface.

[18] Venturi, *RSI* xci. 383.

[19] Ibid. 80–1.

listing pride, envy, glory, honour and shame, riches, religion, women, music, and so on.[20]

His enquiry into their causes and effects has a clear practical purpose: by using the right approach, the general can control and manipulate the human material at his disposal. By 'offering such motives to the troops as naturally tend to raise their courage when depressed and check it when violent or insolent . . . he becomes entirely master of their inclinations and disposes of their forces with unlimited authority'.[21] Echoing de Saxe, Lloyd calls this 'the most difficult and sublime part of this, or of any other profession'.[22]

Montesquieu was the chief inspiration behind another of Lloyd's contributions to the military school of the Enlightenment. Like Guibert, but independently, Lloyd applied Montesquieu's major legacy to the military field. Already in the introduction to the first volume of his *History* (1766), he had emphasized the significance of 'natural history' and 'political law' in determining the face of war. Population, climate, production, soil, government, and similar factors were responsible for the varying national character of the European armies, each of which was briefly discussed by Lloyd.[23] As mentioned above, the 'Essai philosophique sur les gouvernements' was also inspired by Montesquieu, and so were two later works of Lloyd, which were published anonymously and which were presented as parts of the planned treatise 'on the different governments'.

In *An Essay on the English Constitution* (1770), a political pamphlet, Lloyd elaborated on the balance of power in England between monarchy, aristocracy, and democracy. He also proposed a pioneering economic and demographic analysis of military power according to which the size of the population plus the volume of revenues provided a measurement for the 'constant power', the infrastructure of a state.[24] This was a product of his interest in political economy. In *An Essay on the Theory of Money* (1771), in which he advocated the extensive use of paper money, Lloyd

[20] Venturi, *RSI* xci. 69–96.
[21] Ibid. 70.
[22] Ibid. 70.
[23] *History*, vol. i, pp. xxxi ff.
[24] Lloyd, *English Constitution*, ch. IX. Also see id., *An Essay on the Theory of Money* (London, 1771), ch. V; and id., *Rhapsody of the Present System of French Politics* (London, 1779), ch. II.

elaborated on the relationship between the quantity of money in the economy and a series of social measurements.

Simultaneously, Guibert attempted his own application of Montesquieu's legacy to the military field in his *Essai*, which was praised by Lloyd.[25] And in his 'Principles' (1781) Lloyd returned to the same subject in a chapter entitled the 'Connection between the Different Species of Government and Military Operations'. Explicitly relying on Montesquieu, he analysed the military characteristics, institutions, and virtues of despotic, monarchic, republican, and aristocratic regimes.[26]

Interestingly enough, Lloyd's influence on the development of military thought was to transcend his own theoretical intentions and conscious contributions. His history of the Seven Years War was to be instrumental in shifting the interest of military theorists from the organization of armies to the conduct of operations. This field had already been Puységur's main concern and received some attention from all the military thinkers of the French Enlightenment, though they tended to classify it as belonging to the sublime, indeterminant part of the art of war. However, the great successes of the Prussian army were predominantly interpreted in a characteristic structural approach which concentrated on the Prussian military system. Attention thus focused on devising a system of 'tactics' for the French army, and this preoccupation was reinforced even further by the ideas of Folard and Mesnil-Durand that incited the great controversy over the line and the column. Now, Lloyd wrote a widely read campaign history of the greatest war of the period. His controversial account of Frederick the Great's generalship provoked much interest. Attention was beginning to turn from the systems of organization to the conduct of operations.

Puységur wrote that a science of operations had to be based on the study of geography and geometry. From his youth, when he had privately taught geography to officers, Lloyd had earned a reputation for being an expert in this field. His comprehensive survey of British

[25] *History*, vol. i, pt. 2, p. 131.

[26] Ibid. 97–125; the reference to Montesquieu is on p. 98. For an earlier version see the *Rhapsody* (1779), ch. III.

shores from the point of view of a possible invasion was later published in his *Rhapsody*. The *History of the Late War in Germany* (1766) also opened with an extensive geographical survey of the participating states and the theatres of operations. This included a detailed analysis of distances, directions, mountain ranges, river lines, sea-shores, fertility of soil, and density and concentration of population.

This geographical analysis was associated with the growing military use of more accurate maps, made available by the developments in cartography. The advance in this field, stimulated by the great geographical discoveries of the sixteenth century and made possible by the introduction of accurate methods of measurement during the seventeenth century, was enhanced in the eighteenth century by military demands and government involvement. Frederick the Great was still poorly equipped with maps,[27] but, by the second half of the century, most of western and central Europe was covered by an extensive, quite accurate network of maps.

César François Cassini de Thury's thorough topographical survey of France, subsidized by the French government, had begun in the 1730s and was completed in 1789. A fairly accurate topographical atlas of Germany was published in 1750 by the geographical publishing house of J. B. Homann. F. W. Schettan's extensive topographical atlas of Prussia and her neighbours was completed in 1780, but disappeared immediately into the Prussian archives. J. G. A. Jäger's *Grand atlas d'Allemagne* appeared in 1789. The Ordnance Survey was founded in Britain in 1791, concentrating at first on the cartography of the southern counties for military purposes.[28]

Maps not only provided and displayed accurate information on the theatre of operations, but also became increasingly more dominant as the medium of operational planning and staff work. As a result, strategic planning was now commonly thought of in

[27] Christopher Duffy, *The Army of Frederick the Great* (London, 1974), 146–7.

[28] R. V. Tooley and C. Bricker, *A History of Cartography* (London, 1968), 42, 64, 40. Apart from the previous reference and a few words in Colin's *L'Education militaire de Napoléon*, pp. 99–103, there appears to be no study of the development of the military use of maps. What seems to be a very important contribution, Josef Konintz's *Cartography in France 1660–1848* (Chicago, 1987), appeared too late to be consulted in this book.

graphical terms. The movements of armies in space were represented by a whole new range of graphic images. Lloyd introduced one of the first and most useful of these images, the *line of operations*, which represented the communications of the army in the field with its bases of supply, and which expressed one of the dominant features of eighteenth-century warfare.

The ever-increasing size of European armies throughout the modern period, supported by the growing political and financial power of the absolutist state, no longer enabled field-armies to sustain themselves totally on local requisitions of food supplies.[29] On the other hand, the resources of the state also made possible an auxiliary system of supply based on depots and convoys, first organized by Le Tellier and Louvois for the expanding armies of Louis XIV.[30] Coupled with the relative and much stereotyped reluctance of the generals of the *ancien régime* to risk their hard-to-replace troops and political fortunes in a decisive battle, these supply arrangements led to what post-Napoleonic commentators were to call 'wars of manœuvre'. The campaigns of Montecuccoli against the French on the Rhine were among the early examples of this strategic pattern which was dominated by the attempt to threaten the enemy's communications while securing one's own. These practices were therefore already more than a century old when Lloyd introduced the concept of the line of operations in 1781. Old practices were now represented by the new images derived from map planning, and the resulting concepts were to gain dominance because they were perceived as a key for applying the theoretical ideal of the Enlightenment to a new focus of interest, the conduct of operations.

Lloyd introduced the concept of the line of operations only in his theoretical work of 1781 and merely as one among many other themes. He apparently used a concept which was already gaining some currency rather than inventing it himself.[31] He explained that this line, which linked the army in the field to its depots, resulted from the dependence of contemporary European armies on their organized system of supply. The Tartars, for example, who lived

[29] See ch. 1, n. 38 above.

[30] Van Creveld, *Supplying War*, ch. 1.

[31] In his *History*, i. 134, Lloyd writes that 'the line . . . is called The Line of Operations', rather than '*I call* this line . . .'

exclusively off the countryside, were independent of supply lines, and could thus operate with equal freedom in all directions.[32] But this was not the case with modern European armies. The security of their supply line was a central consideration in their operational planning.

From the nature of these lines of operations and supply, several major implications arise. As far as circumstances allow, the shortest and most convenient line must be chosen. It must be directed so as not to be exposed to flank attacks. If extended too far, it might be cut off, leaving the army in the field without supplies in the midst of hostile territory. Thus in order to shorten his lines, the attacker must try to advance his bases as far as possible. On the other side, the defender should manœuvre to threaten the enemy's line of operations, thus forcing him to retreat without even being defeated in battle. The fate of the entire war is therefore dependent on the choice of the line of operations. Other conditions being equal, he who possesses the shorter and more secure lines of operations has the advantage.[33]

Lloyd's *History*, which was very critical of Frederick the Great's generalship in the Seven Years War, provoked in turn much criticism in Germany and Prussia, partly on national grounds, but chiefly for more substantial reasons. Lloyd was clearly biased toward the Austrian cause, and his criticism was often superficial and unsubstantiated. In his Prussian 'counter-history', *Geschichte des siebenjährigen Krieges* (6 vols.; 1785–1801), Colonel, later General, Tempelhoff made this point.

However, while taking issue with Lloyd's interpretation of the war, Tempelhoff, one of the principal military *Aufklärer*s shared his fundamental outlook on military theory and also accepted his reasoning concerning the line of operations. Theory, he explained, was the counterpart of experience in the study of war. Rather than being pedantic, as many regarded it to be, its principles were derived from, and directed towards, action. Without it, everything appeared coincidental, and no analysis was possible in so important and complex a science. In the light of theory, prejudices, errors, and old habits could be rejected.[34] As for the line of operations, armies indeed

[32] Lloyd, *History*, i. 133. [33] Ibid. 134–43.

[34] Georg Friedrich von Tempelhoff, *History of the Seven Years War* (London, 1793), i. 81–2; since the German original was not available to me, references are made to this abridged English edn.

marched on their stomachs; if the need of supplies was not satisfied there could be no operations. In the Thirty Years War it had still been possible to march in any direction, but the larger armies of later times had to rely on magazines, supply convoys, and lines of operations. These lines had to be as short and straight as possible. The success of a campaign was totally dependent upon their security.[35] Now if Lloyd's principles were correct, his application of them had to be wrong; his criticism of Frederick the Great's conduct of operations, written in 1766, was contradicted by his observations on the line of operations developed in 1781.[36]

Lloyd's concept of the line of operations and its broader implications which may be called the 'rationale of operations', were extremely fertile theoretical devices that gave conceptual representation to fundamental features of contemporary and later warfare, and, as such, were to have a long career. This fact has been somewhat obscured by the double-edged revolution in military thinking that was to take place at the turn of the eighteenth century, and whose implications and legacy were extremely hostile towards Lloyd's military ideas.

First, there was the emergence of all-out war, the product of the moral energies and material resources introduced by Revolutionary France. In his rationale of operations based on the logic of supplies, Lloyd reflected the Austrian attitudes to warfare during and after the Seven Years War. This was by far the most extreme example of the alleged reluctance of the generals of the *ancien régime* to risk a decisive battle. To those who experienced Revolutionary and Napoleonic warfare, these traditional attitudes appeared as a gross, absurd error. Indeed, 'absurdity' was the verdict of Napoleon himself on Lloyd's approach to warfare.[37]

The new military outlook was shared by Clausewitz, who, as we shall see, as part of a general reaction against the Enlightenment, also led an intellectual revolution against the traditional conception of military theory. His criticism was fuelled by the tendency, headed by Tempelhoff, to perfect the logic of supply into a more complete and precise rationale of operations, leading to increasingly artificial forms.[38]

[35] Ibid. 61–74. [36] Ibid. 74–81.

[37] Napoleon I, *Notes inédites de L'Empereur Napoléon Ier sur les mémoires militaires du géneral Lloyd* (Bordeaux, 1901).

[38] See e.g. Tempelhoff, *History*, i. 63–74, and his essay on convoys in *History*, ii. 215–36. For Scharnhorst's criticism see p. 164 below; and for Clausewitz's attitude, Clausewitz, *On War*, II, 2, p. 135. For a modern criticism of Tempelhoff's largely artificial portrayal of the contemporary supply system, see van Creveld, *Supplying War*, p. 29.

No one reflected the diversity of ideas and the many conflicts created by this double-edged revolution in military thought more strikingly and extremely than Adam Heinrich Dietrich von Bülow with his extraordinary mixture of old and new. Bülow also completed the shift of interest within the military school of the Enlightenment from the construction of armies to the conduct of operations. Whereas the concept of the line of operations had been only one of the themes in Lloyd's theoretical work, and not even the principal one, Bülow now transformed it into the centrepiece of a new science of operations.

III BÜLOW: BETWEEN A GEOMETRICAL SCIENCE OF STRATEGY AND THE REVOLUTION IN WAR

Renowned as an extremely arrogant and provocative man and working at a time when warfare was revolutionized, Adam Heinrich Dietrich von Bülow (1757–1807) gave the most sensational and controversial expression to each of the changing and often conflicting themes that he propounded during the seven years of his short career as a military theorist. While offering a thoroughly geometrical science of strategy and pushing some of the theoretical notions of the military thinkers of the Enlightenment to the extreme, he was also the most radical advocate both of the old 'war of manœuvre' and the tactical innovations and social resources introduced into war by the Revolution and Napoleon.

Bülow was born in 1757 and joined the Prussian army at the age of fifteen, serving first in the infantry and later in the cavalry. In 1790 he left the Prussian service as a lieutenant and began travelling, trying his luck in unsuccessful commercial enterprises, journalism, and writing. He lived in France, the Netherlands, and England, visited the United States, and wrote his first book on the new American republic (1797). More than a dozen books, primarily on military subjects, followed in less than ten years.[1] In his second book, *Geist des neuern Kriegssystems* (1799), he developed his well-known conception of operations which aroused both positive and negative responses.

Bülow's rationale of operations derived directly from the theoretical and historical reasoning propounded by Lloyd and Tempelhoff.[2] The modern conduct of war, he argued, was based on lines of operations, themselves a product of the greatest revolution in the history of war, the introduction of firearms. Firearms raised

[1] Edward Bülow, a son-in-law of Bülow's brother, General Bülow-Dennewitz who distinguished himself in the Napoleonic Wars, wrote a biographical introduction to a selection from Bülow's writings that he edited in collaboration with Wilhelm Rüstow: *Militärische und vermischte Schriften* (Leipzig, 1853), 3–48. Also see Jähns, *Kriegswissenschaften*, pp. 2133–45.

[2] Bülow himself ignored Lloyd's pioneering role in introducing the concept of the line of operations, and attributed it to his rival, Tempelhoff: see Bülow, *The Spirit of the Modern System of War* (London, 1806), 245. Bülow apparently intended to flatter Tempelhoff who was still alive, active, and close by, and who indeed praised his work.

the need for a regular supply of ammunition. Furthermore, the volume of fire took the place of individual valour as the decisive factor in battle, and consequently the number of troops that a state could throw into war became the determinant factor of military might. According to Bülow's interpretation of the 'military revolution', this, in turn, drove the European powers continually to expand their military forces, thus creating the need for an elaborate supply system—the 'base' of magazines and the 'line of operations' for the movement of convoys to the army in the field.[3] From the campaigns of Montecuccoli against Turenne, and those of Louis XIV, particularly the War of the Spanish Succession, the new supply system became the rationale behind the conduct of operations. The manœuvres of the field-armies and the complex systems of fortresses became the principal means of threatening the enemy's lines of operations while securing one's own, and replaced battle as the centre of warfare.[4]

However, Lloyd's and Tempelhoff's rationale of supply and operations appeared to suggest a wider and more sophisticated theoretical treatment. Could not the armies' movements in space in relation to each other's location and bases be represented geometrically? Puységur had suggested that geometrical study was needed in order to establish the conduct of operations on solid theoretical grounds; and in Turpin's idea to adapt Vauban's siege-system to field warfare there had also been implicit a clear geometrical element.[5] On the whole, Bülow's attempt corresponded to some of the deepest, yet never pursued notions of the military thinkers of the Enlightenment.

An attacking army advancing into enemy territory and towards its objective, the 'object', creates in its movement a sort of imaginary triangle in whose vertex it stands. It draws its supplies from a system of magazines in its rear, at the base of the triangle (the 'base'), and its supply routes form a segment whose boundaries are the sides of the triangle. The defender's field-armies or forces situated in fortresses may penetrate from these sides toward the rear of the attacking army,

[3] *System*, pp. 1–5.

[4] Ibid., and also p. 236.

[5] It is interesting to note that by 1805, Bülow was aware of Turpin's earlier attempt to systematize the conduct of operations. He concluded his *Lehrsätze* by extensively quoting from Turpin's fifty-year-old work: *Lehrsätze des neuern Krieges, oder reine und angewandte Strategie* (Berlin, 1805), 253–74, esp. 261–4.

threaten to cut off the lines of supply, and force the enemy to retreat. Surely, the defender's ability to approach the attacker's rear without being cut off himself depends on the depth of the attacker's advance and the width of his supply lines; or, in other words, on the shape of the imaginary triangle. The deeper the advance of the attacking army and the narrower his supply base, the shorter the distance that the defender, penetrating from the flank, has to cover in order to cut off the attacker, and the safer his own supply lines. In geometrical terms, the narrower the base of the triangle and the longer the perpendicular to it (that is, the narrower the 'object angle'), the easier it is for the defender to cut off the attacker without being cut off himself (see Figure 2).

Could the point from which the attacker's advance becomes, logistically speaking, insecure, be accurately fixed? Bülow claimed that he succeeded in determining it with geometrical precision and certainty. Having examined a series of lines of operations he concluded that an 'object angle' of 90 degrees was the critical point. An angle narrower than 90 degrees did not allow adequate cover for the attacker's lines of operations; and the narrower it became, the more insecure his advance. On the other side, an angle wider than 90 degrees guaranteed the security of the attacker's lines, for by trying to cut them off, the defender exposed his own

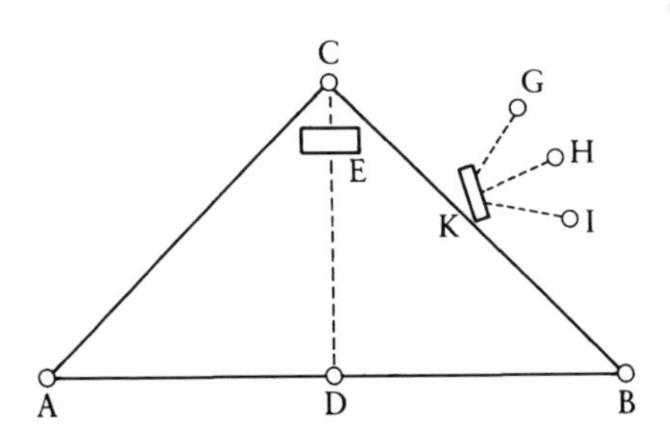

A–B The Base
C The Object
D–C The Attacker's Line of Operations
E The Attacking Army
G, H, I, K The Defender's Fortresses and Field Forces

Fig. 2 A Figure from Bülow's *Spirit of the Modern System of War*

lines; and the wider the angle, the more secure the attacker's advance became.[6]

If indeed this was the case, then Bülow had discovered the mathematical secret of strategy, and established it as a science. From now on, according to Bülow, there was no need for crude considerations and the hazardous trial of battle in order to plan and decide the fate of a campaign. If the attacker relied on an unsound base, the defender could force him to retreat without resorting to battle.[7] Battle was made unnecessary by the scientific perfection of strategy. 'War will be no longer called an art, but a science . . . every one will be then capable of understanding and application; the art itself will be a science, or be lost in it.'[8] The military thinkers of the Enlightenment always left room, alongside the 'scientific' part of war, for that which could not be reduced to rules and principles and was governed by creative genius. Now, 'the sphere of military genius will at last be so narrowed, that a man of talents will no longer be willing to devote himself to this ungrateful trade'.[9]

Indeed, not only the artistic part of war but war itself was to disappear. In the second part of *The Spirit of the Modern System of War*, Bülow analysed the implications of the new military science on the European international system, and reached the conclusion that Europe would be divided into several large states between which perpetual peace would prevail. This was to be the result of the equilibrium inherent in the principle of the base. On the one hand, modern war gave the advantage to large states with mass armies and long borders which provided a wide base. Small states would therefore be swallowed up by the larger ones.[10] There would remain only Spain, France, Italy, Switzerland, the Austrian Empire, Prussia with northern Germany, Denmark, Sweden, Russia, the British Isles, and European Turkey.[11] On the other hand, according to the principle of the base, the strength of the attacker decreased with the increase of distance between himself and his depots. Bülow had already offered a geometrical measurement for the rate of this decrease; now he suggested an arithmetical one, derived directly from Newtonian mechanics. Military force was subjected to the law of gravitation:

[6] Bülow, *System*, 36–68. [7] Ibid. 34–5. [8] Ibid. 228–9.
[9] Ibid. 228. [10] Ibid. 187–97. [11] Ibid. 277–86.

The agency of military energies, like the other effects of nature, becomes weaker . . . in an inverse ratio of the square of the distance; that is to say, in this particular, of the length of the line of operations. Why should not this law, which governs all natural effects, be applicable to war, which now consists in little more than the impulsion and repulsion of physical masses? If, which I do not doubt, it is admissible in the theory of lines of operations, we may in future easily calculate the utmost extent to which military success may be carried.[12]

From a certain point the principle of the base therefore works to the advantage of the defender.[13] 'Every power, then, must ultimately be circumscribed within a certain sphere of military activity, beyond which it must take care not to go.'[14] With the division of Europe between eleven large states, none would be capable of further territorial expansion. War would become pointless. The perpetual peace of the philosophers, propounded shortly before in Kant's *Zum ewigen Frieden* (1795), would be the final result of the principle of the base.[15]

Bülow thus offered not only a geometrical science of strategy but also a mathematical science of politics. Indeed in the twentieth century, he was to be proclaimed a forerunner by the advocates of a geopolitical science.[16]

When Bülow put forward his science of operations in 1799, military practice and theory were already undergoing a far-reaching transformation brought about by the wars of the Revolution and Napoleon. From the late 1790s, a lively debate regarding the significance and scope of this transformation was taking place in professional circles in Germany and raging in the military periodicals

[12] Ibid. 198–9.
[13] Ibid. 213–21.
[14] Ibid. 199.
[15] Ibid. 222–9.
[16] Robert Strausz-Hupé, *Geopolitics: The Struggle for Space and Power* (New York, 1942), 14–21; cited by R. R. Palmer, 'Frederick the Great, Guibert, Bülow' in Earle (ed.), *Makers of Modern Strategy*, p. 69. It is also interesting to note that the surprising similarity of the map of Europe after the unification of Italy and the Austro-Prussian war of 1866 to Bülow's predictions (setting aside the relevance of his analysis) led in 1867 to a republication in Britain of the political part of the *System* as *Pacatus Orbis* (London, 1867).

and literature. Taking the form of a confrontation between the guardians of the old Frederickian system and the supporters of the military as well as social innovations introduced by the Revolution, this debate centred on two main issues: (*a*) the flexible tactics of the French Revolutionary armies, particularly the extensive use of *tirailleurs*, skirmishers in open formation, as opposed to the Frederickian rigid linear tactics; (*b*) the French armies of mass conscription motivated by patriotic and ideological sentiments, as opposed to the professional standing armies of the *ancien régime*, held together by a combination of brutal discipline and *esprit de corps*.[17]

Bülow was fast becoming one of the major advocates of the new Revolutionary warfare and the most provocative critic of the Frederickian system. In *The Spirit of the New System of War* he made the case for the *tirailleurs*, which he elaborated and presented even more forcefully in his later *Neue Taktik der Neuern, wie sie seyn sollte* (1805).[18] This development in tactics had no bearing on his system, but other principal features of the new warfare certainly had. This fact was already becoming manifest in Bülow's own analysis of the campaign of 1800.

In his introduction to *The Campaign of 1800*, Bülow reiterated his claim to be the founder of military science, reasserted the system of the base and the angle of 90 degrees, and made an effort to present them as the rationale behind the French success.[19] However, in the book itself, he hardly resorted to his system. Instead, his analysis concentrated on the social and political infrastructure of Revolutionary France as the major reason for her victory over the Austrian Empire. *Militarily and Politically Considered* was the subtitle of the book. The campaign could only be understood by looking at the nature of the nations involved. The Revolution had abolished feudalism and provided France with mass armies, many times larger than those her enemies could raise, and these were animated by a new

[17] For French Revolutionary tactics see John A. Lynn's new study, *The Bayonets of the Republic* (Chicago, 1984). For the military debate in Germany see: Paret, *Yorck*; Höhn, *Revolution, Heer, Kriegsbild*; W. Shanahan, *Prussian Military Reforms 1786–1813* (New York, 1945), ch. III; and the chs. on Berenhorst and Scharnhorst below.

[18] Bülow, *System*, esp. pp. 174–86.

[19] Bülow, *Der Feldzug von 1800, militärisch-politisch betrachtet* (Berlin, 1801), esp. pp. ix and xiv.

spirit.[20] Money motivated the armies of the *ancien régime*, whereas the Revolution promoted moral forces.[21] Human masses and moral energies were at the root of French power.

The primacy of the social and political infrastructure did not necessarily conflict with Bülow's rationale of operations, but it certainly revealed its very narrow nature which could hardly support his military and political sciences. Furthermore, the rationale itself also suffered devastating blows. Firstly, the foundation of Bülow's system, the logic of supply, artificial as it may have been, was now completely undermined. As he himself, among many others, was quick to note, the armies of the Revolution were living at the expense of the enemy, both financially and logistically.[22] Napoleon's wide-ranging, lightning campaign at the head of the Army of the Reserve, across the Alps, into the Po valley, and towards the Austrian rear, could not be reconciled with Bülow's logistical assumptions, and even less with the 90 degree angle.

Secondly, and even more damaging for Bülow's system, was the fact that Napoleon, who enjoyed new, vast resources and a more flexible military instrument, placed the decisive battle at the centre of warfare. The destruction of the enemy field-army was the goal on which operations focused with a massive and rapid concentration of maximum forces. Once his armies were destroyed, the enemy had to sue for peace. The decisiveness of Napoleon's campaigns struck Europe, adding to the overall picture of the collapse of eighteenth-century warfare. Soon after the appearance of Bülow's system, one of its principal features was thus being theoretically discredited as a result of the revolution in warfare. The following ideas from *The Spirit of the Modern System of War* now stood in stark contrast to the spirit of Napoleon's modern system of war: 'Lines of operations are always directed . . . against the enemy's country . . . and not against the enemy himself; for, the object of war at present should much rather be those places which contain the means of an adversary military power, than men.'[23] And in an even more embarrassing formulation: 'It is more conformable to the genius of war and the latest mode of carrying it out, that a general should make his own magazines and the safety of his lines of convoy, the principal object of his operations, rather than the army of the enemy itself.'[24] Indeed, 'it is always possible to avoid a battle'.[25]

[20] Ibid. 4–8.
[21] Ibid. 5.
[22] Ibid. 5.
[23] *System*, p. 18.
[24] Ibid. 81.
[25] Ibid. 184.

Bülow therefore needed considerable intellectual twists and turns in order to maintain the appearance that his system of operations was perfectly compatible with Napoleonic warfare. He could find support in the fact that the manœuvre against the enemy rear, the *manœuvre sur les derrières*, was the most decisive pattern of Napoleonic warfare, the one used in the campaign of 1800 in Italy.[26] It could be argued that Marengo was merely the inevitable outcome of the envelopment of the Austrian army; the great strategic manœuvre, not the battle which Napoleon nearly lost, decided the fate of the campaign. Indeed, this was the point that Bülow was now to emphasize. His system, he said, placed the manœuvre against the enemy's flanks and rear at the centre of the art of war. This manœuvre aimed at achieving such a strategic advantage that victory would be assured before, or even without, battle.[27] He argued that in *The Spirit of the Modern System of War* he had already stressed the dominant significance of movement and warned against passivity.[28] 'It is a universal law that movement multiplies force.'[29]

The tensions between Bülow's ideas in *The Spirit of the Modern System of War* and *The Campaign of 1800* resurfaced even more forcefully in his two later major works. In his *Lehrsätze des neuern Krieges, oder reine und angewandte Strategie aus dem Geist des neuern Kriegssystems*, published in 1805 but written before the campaign of that year, Bülow again presented his system of 1799, and attempted to demonstrate it, using examples taken chiefly from the campaign of 1800. Some changes were introduced into this new version of the *System*. Firstly, Bülow's geopolitical system, which had hardly progressed towards realization between 1799 and 1805, was excluded from the book. Secondly, the formal certainty of geometry was offered not only for the rationale of operations itself but also for the presentation of the system as a whole. The book was based on three premises from which the entire system was deduced as theorems.[30] However, with the great French

[26] See H. Camon's classical analysis: *La Guerre Napoléonienne*, 2 (Paris, 1907), 9–139.

[27] Bülow, *Der Feldzug von 1800*, p. xii.

[28] Ibid., p. xi.

[29] Ibid. 10–11, 18.

[30] In adopting this Spinozist form, Bülow may have been influenced by J. G. J. Venturini's *Lehrbuch der angewandten Taktik, oder eigentlichen Kriegswissenschaft* (Schleswig, 1800), which was also built in a deductive, semi-geometrical form, though with quite conventional contents. See the following paragraph for Bülow's joint venture with Venturini.

victories of 1805, Bülow returned to the forms of analysis that he had used in 1800–1, being now even more radical, both militarily and politically.

After the successes of his first books, Bülow hoped to obtain a suitable appointment in the Prussian service, but no such appointment was offered to him. He therefore worked as a journalist in London and Paris, where he wrote a book on Napoleon, *Über Napoleon Kaiser der Franzosen* (1804).[31] Later, in 1806, in Berlin, he co-edited a military journal entitled *Annales des Krieges*.[32] The other editors included Venturini, J. Voss, and another celebrated military thinker, Georg Heinrich von Berenhorst, the most respected critic of the Frederickian system, who greatly influenced Bülow in this respect, despite the paradigmatical gulf between their conceptions of military theory.[33] Bülow's criticism of his country, blended with personal frustration, extreme self-esteem, and a provocative style, became bitingly sarcastic in response to the collapse of the powers of the *ancien régime* in the campaign of 1805.

Within three months, a gigantic campaign and two decisive military encounters brought about the military destruction and virtual occupation of the Austrian Empire, which was supported by the armies of Russia. Such a fate had befallen no major European power in the modern period. The traditional European balance of power broke down, and Prussia found herself exposed and in an extremely dangerous diplomatic and military position. At this moment of crisis, Bülow wrote *The Campaign of 1805, Militarily and Politically Considered*, which was published, because of its radical ideas, at the author's own expense. Heterogeneous in composition, the book combined a description of the campaign with a political and military programme for the transformation of the Prussian state. It censured the Prussian system in the name of the new political and social order of Revolutionary France.

A comparison between the states of Germany and France revealed the former's inferiority in terms of the socio-political infrastructure. Bülow alleged that in order to survive, the Prussian state must undergo comprehensive reforms. She must give priority to talent over birth, and make full use of social potential by introducing general conscription and opening her administration and officer corps to the

[31] Jähns, *Kreigswissenschaften*, p. 2133.

[32] Bülow and Rüstow (edd.), *Schriften*, p. 24.

[33] See Ch. 5. II below.

able. The system of social rewards must support this aim by promoting utility to the state. Imitating the model of the French Legion of Honour, Bülow proposed an elaborate scheme for three orders of merit.[34]

In the strictly military field, Bülow faced his old dilemma. He had to reconcile Napoleonic warfare with the conceptions on which his reputation rested. At Ulm Napoleon brought the strategic manœuvre against the enemy's rear to its pinnacle. Mack's army, more than 70,000 men strong, capitulated without battle after being placed in a hopeless strategical position. The *Grande Armèe* that marched from its long and enveloping lines along the Rhine and the Main, cut it off from Austria in a sweeping movement.[35] Ulm, like Marengo, could therefore be presented as consistent with Bülow's system of operations of 1799; again, the strategic manœuvre overshadowed, and even eliminated, the need for battle. However, it was difficult to explain Austerlitz in this manner. Furthermore, as in the campaign of 1800 in Italy, the encirclement was achieved not on the basis of the system's calculations. Bülow's own account clearly expressed the real secret of the Napoleonic conduct of operations: the emperor 'uses his capital'.[36] A revolutionary exploitation of initiative, mobility, and concentration of force is responsible for his success. He executes the doctrines of Guibert.[37]

Bülow's writings were more than the Prussian government was prepared to tolerate. When Prussia faced her gravest trial, Bülow described Austerlitz as the modern Actium and predicted a French hegemony over Europe.[38] He was arrested, declared insane, and detained first in Berlin, and later, with the fall of the city and the French advance, in Colberg and Riga under Russian custody. In 1807 he died in prison, according to his relatives, due to ill-treatment.[39]

Bülow's novel, sensational, and controversial works attracted wide attention, and made his name known throughout Europe. The *System* was republished in 1805 (and again in 1835), and was translated into French (1802, reprinted 1814) and English (1806, reprinted 1814, 1825). *The Campaign of 1800* was translated into

[34] Bülow, *Der Feldzug von 1805, militärisch-politisch betrachtet* (n.p., 1806), vol. ii, pp. xviii–xxxiii, 108, 132 ff.

[35] Ibid., vol. i, pp. lxiii and lix.

[36] Ibid., vol. ii, p. 109.

[37] Ibid., vol. i, p. lix.

[38] Ibid., vol. ii, 158.

[39] Bülow and Rustow (edd.), *Schriften*, pp. 37 ff.

French in 1804.[40] Reactions were numerous and polarized. Bülow referred extensively to them in his works and even included large sections from his critics' comments on various subjects in his *New Tactics of the Moderns as They Should Be*.[41]

Regarding Bülow's system, several characteristic approaches can be discerned. Tempelhoff, who in the last years of the century headed a tendency to analyse the conduct of operations according to an increasingly formal rationale of supplies, praised Bülow's work which could be seen as the logical conclusion of his own.[42] Likewise, General Binzer, the Chief of Staff of the Danish army wrote a short and very complimentary book on Bülow's work.[43]

However, it is important to understand that while Bülow attempted to realize some of the deep-rooted but remote theoretical ideals of the military thinkers of the Enlightenment, he, *ipso facto*, violated the tenets of their traditional, well-established theoretical outlook. In his introduction to the English translation of the *System* (1806), Malorti de Martemont expressed this clearly. The art of war, he wrote, would never become totally scientific as Bülow suggested. While in part it could be reduced to rules and principles, another part, influenced by the diversity of political, moral, and physical conditions, was perpetually wavering, and required application by creative genius.[44] Scharnhorst, one of the most distinguished military *Aufklärers*, who rejected the new theoretical trends, criticized Bülow's career in his book review 'H. v. Bülow nach seiner Hypergenialität und seinen Abenteurn geschildert'.[45] Finally, under Scharnhorst's influence at the Institute for Young Officers in Berlin, the young Clausewitz developed a new theoretical outlook in a double-edged reaction against the theoretical legacy of the Enlightenment and the war of manœuvre.

[40] An edn. with Napoleon's notes written at St Helena appeared in 1831; a later edn. was published in 1841.

[41] Bülow, *Neue Taktik der Neuern, wie Sie seyn sollte* (Leipzig, 1805), 175–300.

[42] Jähns, *Kriegswissenschaften*, 2142.

[43] J. L. J. Binzer, *Über die militärischen Werke des Herrn von Bülow* (Kiel, 1803; repr. 1831).

[44] *System*, pp. iii–vii.

[45] Published in the *Göttinger gelehrten* (Berlin, 1807), Jähns, *Kriegswissenschaften*, p. 2142; for Scharnhorst's theoretical position, see Ch. 6. I below.

Clausewitz's critique of Bülow's *Lehrsätze des neuern Krieges* was published anonymously in 1805 in the military periodical *Neue Bellona* under the title 'Remarks on the Pure and Applied Strategy of Mr von Bülow'. The twenty-five-year-old officer, who admitted that in his youth he had been attracted by Bülow's vision, now sharply criticized his theoretical outlook and system.[46] First, from Clausewitz's point of view, determined by his particular interpretation of Napoleonic warfare, Bülow's promotion of the strategic manœuvre and rejection of battle amounted to a totally false conception of the nature of war. Bülow refused to understand what the whole world had already learnt to accept—that tactics was about fighting and centred on the engagement.[47] Bülow's geometrical system was equally false and artificial, and therefore constantly conflicted with reality. The tension that emerged in Bülow's own works did not escape Clausewitz; Bülow's own examples refuted his principles.[48] All this was the unavoidable result of the attempt to force a priori mathematical categories on the diversity of historical experience; Bülow lacked a critical historical approach.[49] Not only was history adapted by Bülow to fit his theory but everything that was not consistent with his desire to systematize was ignored. He focused on the geographical factors because they lent themselves to quantitative analysis, but disregarded the nature of the people involved, the moral forces that animate war, and the enemy against whom the war was directed. War was a map-game for him.[50] A true study of war must take into account the full diversity and complexity of the conditions involved.[51] Bülow's system was but one abstraction on top of the other; a single concept was generalized to create a false science.[52]

In *On War*, Clausewitz repeated his early criticism:

> One ingenious mind sought to condense a whole array of factors, some of which did indeed stand in intellectual relation to one another, into a single concept, that of the *base* . . . He started by substituting this concept for all these individual factors; next substituting the area or extent of this base for the concept itself, and ended up by substituting for this area the angle which the fighting forces created with their base line. All this led to a

[46] Clausewitz, 'Bemerkungen über die reine und angewandte Strategie des Herrn von Bülow', *Neue Bellona*, IX 3 (1805); repr. in *Verstreute kleine Schriften*, W. Hahlweg (ed.), (Osnabrück, 1979); see p. 87.

[47] Ibid. 70, 78–9. [48] Ibid. 75–6, 84–7. [49] Ibid. 87.

[50] Ibid. 73, 79, 81. [51] Ibid. 82. [52] Ibid. 87.

purely geometrical result, which is completely useless. This uselessness is actually inevitable in view of the fact that none of these substitutions could be made without doing violence to the facts and without dropping part of the content of the original idea. The concept of a base is a necessary tool in strategy and the author deserves credit for having discovered it; but it is completely inadmissible to use in the manner described.[53]

In the criticism of Bülow's theoretical outlook one point has remained unnoticed: the geometrical basis of the system itself was simply wrong. Bülow's rationale of operations is based on the fact that the attacker's lines of supply in the midst of hostile territory rely on a much narrower base than that of the defender who operates in his own territory. In a geometrical formulation: the attacker draws his supplies from a narrow, triangular segment of space smaller than 180 degrees, whereas the defender can draw his from all the rest of the space's circumference (see Fig. 3). This situation gives the defender a clear advantage in a contest of manœuvre whose aim is to cut off the enemy's lines while preserving one's own.

Let us examine the situation described by Bülow.[54] In order to place himself at the attacker's rear, the defender (D) has to cover a shorter distance than the one that the attacker (A) must cross in attempting a counter-move (Fig. 4). Therefore in the event of such a move, the defender can withdraw to cover his own lines and still retain his threatening position at the attacker's rear. Now, his advantage is decreasing in proportion to the segment of space on which the attacker relies; the distance that he must cross in order to place himself at the attacker's rear increases and the attacker's prospects of carrying out a counter-manœuvre improve.

However, the 90 degree angle is not, as Bülow suggests, a turning-point from which the attacker's route becomes shorter. Even when the 'objective angle' becomes obtuse, the attacker still has a longer distance to cover in his counter-manœuvre, because he moves on the longest side of a triangle (Fig. 5). Thus, though the defender's advantage decreases with the increase in the attacker's angle, it does not disappear until both parties rely on a similar segment of space, that is until the attacker's salient disappears. Then,

[53] Clausewitz, *On War*, II, 2, p. 135.

[54] *System*, pp. 38–9.

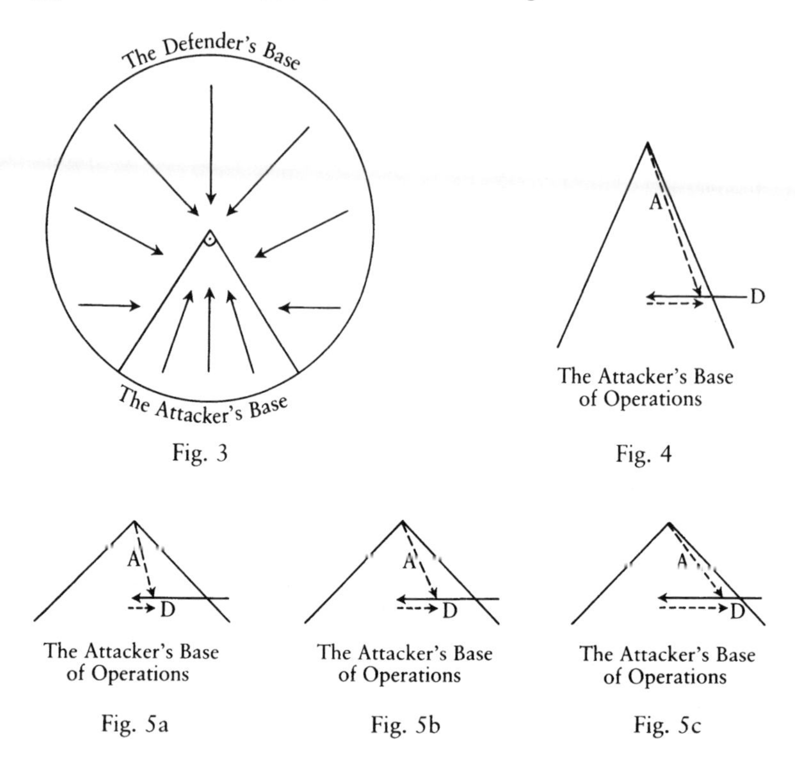

Fig. 3

Fig. 4

Fig. 5a

Fig. 5b

Fig. 5c

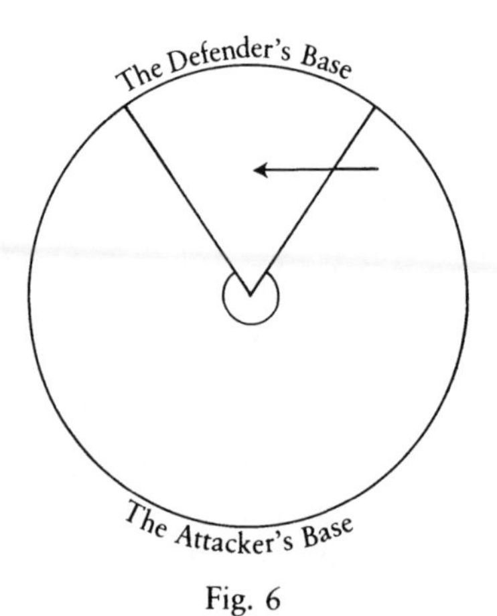

Fig. 6

Figs. 3–6 The Geometrical Rationale of Bülow's System

if the attacker operates from a base larger than 180 degrees, he himself enjoys, as Bülow points out, the advantage of a wider base (Fig. 6).[55]

All the exciting conclusions drawn from the 90 degree angle were therefore without foundation. The rationale of supplies and manœuvre amounts to no more than the obvious: that salients are exposed to being cut off.

The error in the geometrical core of Bülow's system of operations is, of course, no more than a curiosity. It simply demonstrates that even here Bülow's thinking was sloppy and superficial. Unfortunately, Bülow's sensational geometrical system, coupled with the obscurity into which the military school of the Enlightenment has sunk in historical consciousness, have led to a remarkably vague and stereotyped image of Clausewitz's predecessors. This consisted of the largely mythical post-Napoleonic trio of the eighteenth-century geometrical school, the war of manœuvre, and military conservatism.

Firstly, there was no real 'geometrical military school' in the eighteenth century. As we have seen, there were certainly deep-rooted geometrical notions and a remote ideal. Linear tactics also encouraged the extensive use of graphic schematizations but, apart from rare exceptions, they were hardly more than illustrations.[56] Finally, there was the search for the rationale of operations which emerged towards the end of the century, and which only Bülow developed into a geometrical system of operations. What can be described as a geometrical school of operations existed, in fact, to some degree only in the Napoleonic period, when military writers such as August Wagner and, more importantly, Archduke Charles, who were influenced by Bülow, based their analysis of operations on geometrical forms and considerations, though without Bülow's claims to mathematical precision and rigour.

In any case, the geometrical attempt was by no means inseparably linked with the idea of the rationale of operations. Bülow's fantastic system must not obscure this. Bülow's work attracted lively interest

[55] *System*, p. 65. Indeed the 90 degree angle leads to a paradox; the attacker is supposed to have the advantage when he relies on more than a 90 degree angle; but at the same time, and until the attacker relies on one of more than 270 degrees, the defender too relies on more than 90 degrees and should have the advantage himself.

[56] For F. Miller's *Reine Taktik* (Stuttgart, 1787–8), see also p. 164 below.

because it corresponded to a widely held feeling that the relative movement of contemporary armies in the theatre of operations, in relation to each other's position, communications, and objectives, was susceptible to a fruitful schematization in terms of time and space. This was by no means an unsound view. Lloyd had already proffered a quite penetrating analysis of the war of manœuvre which, since the Seven Years War, was gaining favour among the Austrian high command. But the military thinkers of the Enlightenment were far from being universally identified with the war of manœuvre, and it is enough to cite Guibert in this context. Indeed, now, the rationale of operations required a new formulation in terms of Napoleonic warfare.

The many different intellectual, military, and political transformations at the turn of the nineteenth century were therefore reflected in a variety of individual expressions. Bülow combined an extreme statement of some of the theoretical notions of the Enlightenment with the strategy of manœuvre, and military and political radicalism. Archduke Charles combined the theoretical outlook of the Enlightenment with a conservative adaptation of the strategy of manœuvre and the military and political institutions of the *ancien régime* to the Napoleonic era. Jomini synthesized the theoretical legacy of the Enlightenment with Napoleonic warfare, developing an updated, credible, and highly successful rationale of operations. Berenhorst expressed a Counter-Enlightenment point of view and harsh criticism of the Frederickian political and military system. Scharnhorst fused the classical views of the *Aufklärung* with reformist political and military positions. And Clausewitz combined his political and military reformism with the intellectual outlook of the German Movement and the Napoleonic war of destruction.

4
Through the Napoleonic Age

I ARCHDUKE CHARLES AND THE AUSTRIAN MILITARY SCHOOL

The image of the eighteenth century as an era of limited political aims and cautious strategy of manœuvre—an image created by the men of the post-Napoleonic period and highlighted by the German military school of the nineteenth century—is somewhat stereotyped. It is true that compared with the age of the wars of religion or the age of national wars, the wars between 1648 and 1789 were indeed relatively limited in their scope and aims. As has been progressively recognized since the days of Delbrück, the lack of ideological fervour, coupled with the *Realpolitik*, the restrictive social structure, and the professional armies of the *ancien régime* were all responsible for this. However, politically, the successive coalition wars against Louis XIV, Maria Theresa, and Frederick the Great involved not only a heavy strain on the resources of the countries of those monarchs, but also (particularly in the latter cases) the prospect of grim political consequences in the event of defeat. As to the military aspect, the campaigns of Marlborough or Frederick, which between them dominated eighteenth-century warfare, were hardly characterized by an unwillingness to fight. Nor did the French or the Austrians (the latter, at least in the first half of the century) shrink from major battles.[1]

If the eighteenth century came to be so strongly identified as the era of manœuvre warfare, it was predominantly because of tendencies which had become increasingly prominent late in the century, first in Austria and then in Prussia, and which were violently challenged with the coming of the Revolution and Napoleon. After the traumatic experience of the War of the Austrian Succession, and in the face of the superior qualities of the Prussian army and the military genius

[1] See C. Duffy, 'The Seven Years War as a Limited War', in Rothenberg *et al.* (edd.), *War and Society*, pp. 67–74.

of Frederick, the Austrians adopted a cautious strategy. Like the Dutch military school in the protracted wars against Spain and France from the late sixteenth to the early eighteenth centuries, the Austrian armies made extensive use of strong defensive positions, field-works, and fortresses, and fully exploited the leverage of supply and communications, rather than risk an open battle. Indeed, shaken by his own ordeal in the Seven Years War, alarmed by the increasing human cost incurred in order to drive the Austrian armies out of their positions, and concerned by the growing number of fortresses, Frederick himself in the last decades of his reign moved away from the lightning strategy of his great wars, which he no longer considered feasible. He made this clear in his *Militärische Testament* (1768), and conducted a campaign of positions in the diplomatic and bloodless War of the Bavarian Succession (1778–9). These attitudes, fully reflected in the theoretical works of Lloyd and Tempelhoff, thus became prevalent in the Austrian and Prussian armies when they encountered Revolutionary France.

After participating half-heartedy in the early campaigns of the first coalition (1792–5), Prussia did not return to the war against France until 1806, when her army was destroyed by Napoleon's mass armies and crushing strategy. On the brink of destruction, Prussia had to adapt to the new character of war, embarking on inseparable political and military reforms, laying the foundations for a national army, and adopting an active, battle-oriented strategy. Events were different, however, in Austria. Though forced to reform her military organization during her long and intermittent struggle with Revolutionary France and Napoleon, Austria was far less susceptible of change than even the Prussia of the *ancien régime*. The heterogeneous character of her political structure and particularly her deep ethnic fragmentation, placed Austria in a state of fundamental disadvantage in the age of national war.[2]

Hence the closely linked themes in the distinctive approach of the Austrian school to war. The mobilization of mass armies and popular energies were in conflict with the empire's very *raison d'être*. Limited

[2] See esp. K. Peball, 'Zum Kriegsbild der österreichischen Armee und seiner geschichtlichen Bedeutung in den Kriegen gegen die Französische Revolution und Napoleon I', in W. v. Groote and K. J. Müller (edd.), *Napoleon I und das Militärwesen seiner Zeit* (Freiburg, 1968), 129–82; and G. E. Rothenberg, *Napoleon's Great Adversaries, The Archduke Charles and the Austrian Army, 1792–1814* (London, 1982).

conscription, modelled on the Prussian 'canton system', was introduced in 1771, but attempts to form a second-line militia before the war of 1809 were treated with distrust by Archduke Charles and the Austrian high command, and never took off. In the struggle against the superior Napoleonic power, Austria was therefore totally dependent on her standing army which was large, but expensive and difficult to replace. Safeguarding this army and ensuring that it was not rushed into major battle under less than favourable conditions were thus paramount considerations for both Daun and Charles, even to the point of letting many potentially decisive opportunities slip away. Indeed, despite fierce personal and political rivalry, and considerable differences in temperament and style of generalship between such men as Daun, Lacy, and Loudon in the Seven Years War, or Archduke Charles and Schwarzenberg in the Napoleonic Wars, these general notions and attitudes underlay the Austrian conduct of war.

The widely respected theoretical works of Archduke Charles stand out in the comparatively meagre output of military literature in Austria of the Enlightenment.[3] Like Charles's active but less-than-bold generalship, and comprehensive but pronouncedly limited military reforms, these works are a striking expression of Austria's fundamental condition during the transition from old to new.

Archduke Charles (1771–1847), the son of Emperor Leopold II and the younger brother of Francis I, first experienced war against Revolutionary France in the campaigns of 1793–4 in Flanders. In 1796, he defeated Jourdan and Moreau in an excellent campaign in southern Germany, and although beaten by Napoleon in the Tyrol a year later, he again fought successfully against Jourdan in the German theatre of operations in 1799.

Acknowledged as the best general of the Habsburg monarchy, Charles was called on to reorganize the Austrian army after the defeat of 1800. He was appointed field marshal, president of the *Kriegshofart*, and head of a newly formed ministry of war, thus securing a considerable degree of control over the deeply factional Austrian high command. His brother the emperor, in accordance

[3] See the very sketchy treatment of Manfried Rauchensteiner, 'The Development of War Theories in Austria at the End of the Eighteenth Century', in G. Rothenberg *et al.* (edd.), *War and Society*, 75–82.

with what one historian has called the 'Wallenstein complex' of the Habsburg monarchy, took, however, special care that he did not have a free hand in the army or a say in political matters.[4] The endemic friction with the crown and court, and Charles's lack of sufficient personal authority in the army itself, became more pronounced before the renewed outbreak of hostilities in 1805. Objecting as he did to the war against France, Charles was stripped of some of his authority, but was none the less given command in Italy, the anticipated main theatre of operations. This time, however, Napoleon chose the Danube valley for his main thrust, and Charles was too slow to influence the course of events which culminated at Austerlitz.

After that defeat, Charles regained control over the Austrian army with the rank of generalissimo. Though opposed to the new war with France in 1809, he led the Austrian army to victory at the Battle of Aspern-Essling over the French army headed by Napoleon, who had entered Vienna and was attempting to cross the Danube to the north. The war was decided in favour of the French only six weeks later at the heavy and drawn-out battle of Wagram. After signing an unauthorized armistice with Napoleon, Charles was relieved of all duties. Regarded as too uncontrollable by the emperor and Metternich, he was not recalled during the last campaigns against Napoleon in 1813–15, and never again saw active service.[5]

This fact did not, however, diminish Charles's universal reputation as the best general in continental Europe east of the Rhine, nor the widely held respect for his personality and military record. His historical and theoretical works were therefore received with much interest, particularly his accounts of the campaigns of 1796 and 1799, and his *Principles of Strategy*, which appeared for public distribution

[4] G. Craig, 'Command and Staff Problems in the Austrian Army, 1740–1866', in M. Howard (ed.), *The Theory and Practice of War* (London, 1965), 45–67.

[5] Excluding some half a dozen popular biographies and numerous accounts of his major campaigns, the most comprehensive (official and semi-official) studies of Charles's life, incorporating a great deal of primary material, appeared in the late 19th cent., at the same time as the publication of Charles's collected works. These multi-vol. biographies include: H. R. v. Zeissberg, *Erzherzog Carl von Oesterreich* (Vienna and Leipzig, 1895); M. E. v. Angeli, *Erzherzog Carl von Oesterreich als Feldherr und Heersorganisator* (Vienna and Leipzig, 1896); and O. Criste, *Erzherzog Carl von Oesterreich* (Vienna and Leipzig, 1912). For a modern work in English, see Rothenberg, *Napoleon's Great Adversaries*.

in 1814 and was quickly translated into French (1817) and Italian (1819).[6]

Charles's work follows three well-established theoretical paths: the theoretical outlook of the Enlightenment; the rationale of communications as devised by Lloyd and Tempelhoff and only mildly adapted to Napoleonic strategy; and the geometrical analysis introduced by Bülow, though without his wilder pretensions to scientific and mathematical rigour.

Charles's first major theoretical treatise, *Principles of the Higher Art of War* (1806), was written in collaboration with his military mentor, Colonel Lindenau, once an adjutant of Frederick's and an author on tactics, and with General Mayer, his chief of staff and a close supporter.[7] It was distributed among the generals of the Austrian army as part of a comprehensive vitalization of instruction material. The book concludes:

> The principles of the science of war are few and unchanging. Only their application is never the same and can never be the same. Every change in the conditions of armies: in their arms, strength and position, every new invention, involves a different application of these rules.[8]

Thus, while 'the principles of war are founded on mathematical, evident truths', judgement, trained by historical study and military education, presides over their application. Both the science of principles and historical study must reinforce genius and experience in the making of generals.[9]

In his early work 'On the War with the New Franks' (1795), an analysis of the Austrian conduct of war in the first campaigns against the armies of Revolutionary France who were still ill-organized, Charles was critical of his country's military strategy. He argued that, in view of their enemy's relative weakness, the Austrians were much too defensive, were excessively concerned over the safety of their

[6] Archduke Charles [Carl von Oesterreich], *Ausgewählte Schriften* (*AS*), F. X. Malcher (ed.) (Vienna and Leipzig, 1893–4), include six vols. and an atlas. A concise edn. of his major theoretical works, *Ausgewählte militärische Schriften* (Berlin, 1882), appeared in the *Militärische Klassiker* series.

[7] Rothenberg, *Napoleon's Great Adversaries*, p.106. Karl Freidrich von Lindenau's major work, *Über die höhere preussische Taktik* (Leipzig, 1790), is a programmatic scheme of evolutions in linear formation.

[8] *Grundsätze der höheren Kriegskunst*, *AS* i. 50.

[9] Ibid. 49–51; the same ideas are expressed in Charles's later *Grundsätze der Strategie*, *AS* i. 231–3.

magazines and communications, and dispersed their forces too widely in the strategic deployment devised by Lacy and known as the 'cordon system', thus ending up inferior everywhere.[10]

However, a decade later, when he commanded Austria's war effort against the full might of Napoleonic power, Charles's views no longer diverged from his country's traditional attitudes. Though stressing the principle of the concentration of force at the decisive point—probably under Jomini's influence—[11] he was far from advocating bold action. Dispassionately he wrote that a 'mathematical truth teaches us that a decisive result cannot be achieved when totally equal forces operate against one another'.[12] All the more so, as Austria was not even equal to France, despite the considerable expansion of her standing army. The underlying link between Austria's strategic position and Charles's theoretical conceptions is clear:

> Only when the last object, which is essential for the survival of the state is about to fall into the hands of the enemy, when no other means of relief is left open, may the general risk a battle even with inferior forces; then he may depart from every rule . . . It is a battle of despair, the loss of which one does not survive.[13]

Indeed, for Charles, the focal point of war lay not in battle but elsewhere: 'A principal rule in offensive as well as in defensive war is never to choose with one's main force a line of operations or position in which the enemy is close to our lines of communications and magazines.'[14] The attacker should seek to penetrate into his enemy's country in order to cut him off from the means that support his war effort, whereas the defender must cover his communications and play for time.[15]

The affinity between Charles's cautious political premises and his characteristic strategic outlook is again revealed in *Principles of Strategy*:

> The events of war have such decisive results that it is the general's first duty to secure the outcome as far as he possibly can. But this can only be achieved if the means required for the conduct of war are available . . . Every

[10] 'Über den Krieg mit den Neufranken', *AS* v. 5–15, esp. 6–7.

[11] *Grunsätze der höheren Kriegskunst*, *AS* i. 3–4; Jomini's probable influence was pointed out by H. Ommen, 'Die Kriegsführung des Erzherzogs Carl', *Historische Studien*, XVI (1900), 109.

[12] *Grundsätze der höheren Kriegskunst*, *AS* i. 50.

[13] *Grundsätze der Strategie*, *AS* i. 330.

[14] *Grundsätze der höheren Kriegskunst*, *AS* i. 6.

[15] Ibid. 7–8.

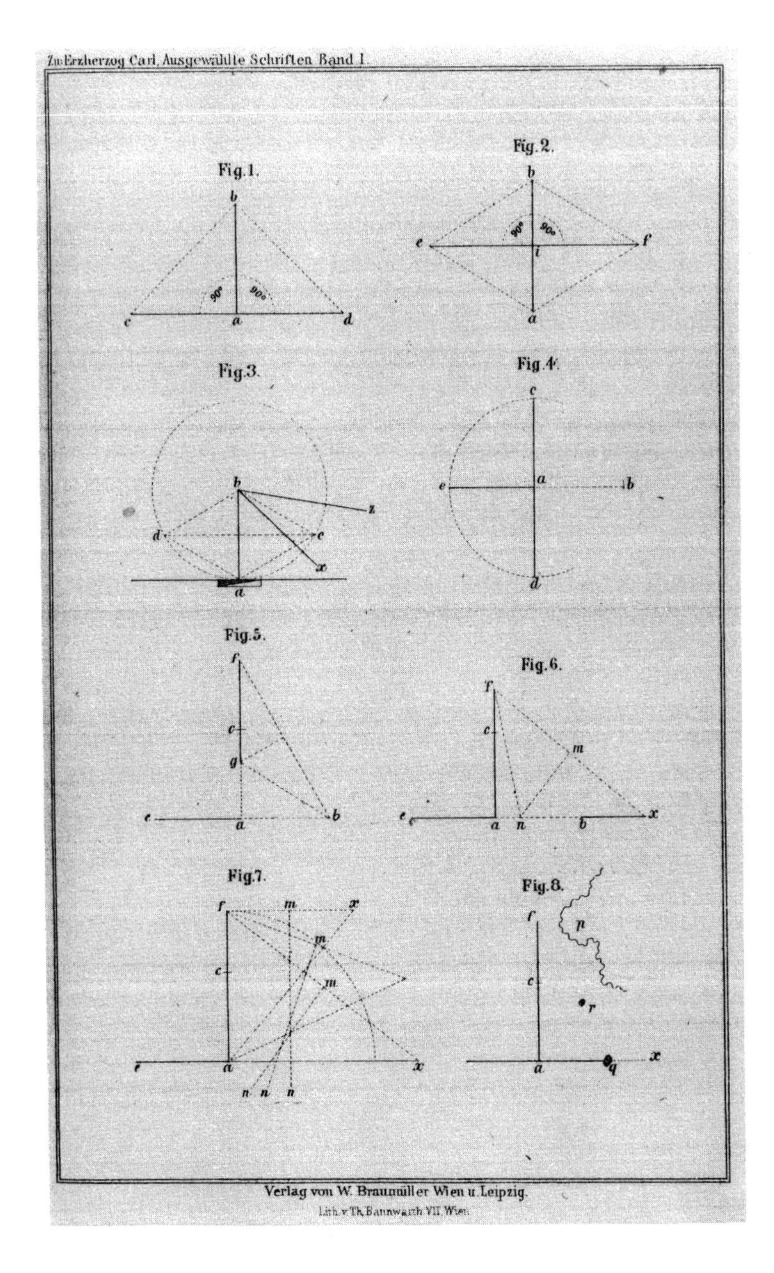

Fig. 7 Geometrical Figures from Archduke Charles's *Principles of Strategy*

deployment and movement must therefore provide full security for the key to the country behind, for the *base of operations* where supplies are accumulated, for the communications with these supplies, and for the *line of operations* chosen by the army for his advance from its base to the *objective of operations.*[16]

Indeed, Charles adopted not only Bülow's general theoretical scheme, but also the geometrical analysis of operations, though without the mathematical centre-piece of Bülow's system—the 90 degrees angle and its exciting implications (see Fig. 7). Thus, while Charles's conclusions regarding the ability of armies to cover their base and line of operations are not as sensational as Bülow's, they do tend to appear somewhat trivial as well as unnecessary.[17] 'On the one hand', wrote Caemmerer, 'they prove in a very roundabout way things quite obvious; and . . . on the other hand, the results remain highly debatable, since in war not only distance has to be considered, but also direction, number and conditions of the road.'[18]

Charles's main contribution to the rationale of operations is the concept of strategic key-points which dominate the base, communications, and objective. Situated at the most vital junctions and channels of movement, these geographical strongholds—particularly in closed regions—constitute the 'key to the country' and to the conduct of a campaign. Their identification and seizure ought therefore to be the general's first consideration in operational planning.[19]

The appearance of Charles's *Principles of Strategy* (1814) and the favourable reception it enjoyed, which could not, naturally, be dissociated from the author's royal status and military prestige, evidently alarmed Jomini, since it threatened to overshadow his own work and undermine his claim to be the founder of military science.

[16] *Grundsätze der Strategie*, *AS* i. 237; my emphases. While Ommen ('Kriegsführung', p. 121) is quite right in pointing out that Charles had already used the concept of the base, related to the system of fortresses and depots, as early as 1795 (*AS* v. 9), Charles's debt to the systematical conceptual framework formulated by Bülow is obvious.

[17] For the geometrical analysis see *Grundsätze der Strategie*, *AS* i. 237–40 and tables.

[18] R. v. Caemmerer, *The Development of Strategical Science during the 19th Century* (London, 1905), 58.

[19] *Grundsätze der Strategie*, *AS* i. 240–3.

However, for obvious reasons, Charles could not be dismissed in the typical Jominian fashion like the scores of military writers mentioned in the *Treatise* and later *Summary*. A more diplomatic approach was called for here. Charles was therefore the only military author to whom Jomini gave, literally, a 'royal treatment', and with whom he was even prepared to 'share' the leadership of military theory, not, however, without stressing his own pioneering position.[20] On Charles's request, he also agreed to take upon himself the translation of *Strategie* into French. As we shall see, Charles's work was partly responsible for Jomini's decision to write the *Summary*, and also left its mark on the character of that book.

Given the nature of Charles's ideas, Clausewitz's treatment of Charles is conspicuously inconspicuous, lacking Clausewitz's usual zeal in dealing with other, less eminent military thinkers whose ideas paralleled those of Charles. In *On War*, referring to Charles as 'a sound historian, a shrewd critic and, what counts even more, a good general', he criticized neither Charles's military outlook nor his geometrical theories which he would normally have treated with hail and thunder.[21] His strongest criticism of Charles's military outlook appears in his work on the campaigns of 1799 in Italy and Switzerland.

> Firstly, he [Charles] lacks an enterprising spirit and the hunger for victory. Secondly . . . while his judgement is generally good, he has fundamentally a completely erroneous view of strategy. In war all should be done in order to destroy the enemy's forces, but this destruction does not exist as a separate aim in his conceptual outlook. . . . For him success is merely the occupation of certain positions and areas.[22]

What Clausewitz wrote when Archduke Charles was at the height of his reputation as the best general of a respected European power, became the prevailing view when Austria declined into a second-rate power, and the German military school, expressing the might of a united Germany, dominated military theory, naming Clausewitz as its forefather, and promoting crushing decision in battle. Caemmerer's review of Charles's theoretical works is characteristic:

[20] Jomini, *Treatise on Grand Military Operations* (New York, 1865; based on the 3rd edn. (1818) of the *Traité*), p. xxi; id., *Summary of the Art of War*, pp. 13–14.

[21] Clausewitz, *On War*, VI, 16. p. 123.

[22] *Die Feldzüge von 1799 in Italien und in der Schweiz*, in Clausewitz, *Hinterlassene Werke*, (Berlin, 1832–7), v. 152.

How strange these [Charles's] words sound if we consider that at the time they were written the man [Napoleon] who so impressively had taught the world the importance of tactical success, was at the zenith of power and glory. In all this [Charles's works] we cannot find a trace of cheerful confidence in one's strength and ability.[23]

Charles's erroneous strategic outlook, wrote Caemmerer, stemmed from, and in turn reinforced, military tendencies which proved catastrophic for the destiny of the Austrian empire. The origins of this outlook were to be found in Field Marshal Daun's headquarters in the Seven Years War, and its influence could be discerned in the extreme caution and peculiar manœuvres of the Austrian army under Schwarzenberg in the campaign of 1814 in France, and again in the wars of 1859 and 1866. Instead of seeking battle, the Austrian generals looked for 'key-points'.[24]

While the Austrian military school certainly tended towards theoretical artificiality and 'strategic mannerism', Caemmerer's judgement is typical of the refusal of the German military school to acknowledge the deeper historical and strategic roots of any military outlook different from its own. This is revealed remarkably by Caemmerer's own conclusion written in 1904:

At the present moment we live in an age where an extraordinary progress in the technics of firearms exposes us to the danger of over-estimating the value of defensive positions, and where such theories of the importance of ground in strategy as advanced by the Archduke Charles might again become that serious danger for weak minds. . . . It was therefore necessary to leave no doubt about their failure in history.[25]

Ironically, within ten years, in encountering new historical conditions, it was the military outlook of the German military school that was shattered and called into question, while 'the value of defensive positions' and 'the importance of ground' became paramount.

[23] Caemmerer, *Strategical Science*, p. 61.

[24] Ibid. 61, 69; for an even more nationalistic example of the German school's attitude, which contrasts Charles and his milieu with the new German spirit from Clausewitz to Bismarck, Nietzsche, and Treitschke, see R. Lorenz, 'Erzherzog Carl als Denker', in A. Faust (ed.), *Das Bild des Krieges im deutschen Denken* (Stuttgart and Berlin, 1941), 235–76.

[25] Caemmerer, *Strategical Science*, p. 70.

Because of the situation in which Austria found herself in fighting with the superior Napoleonic power there was no reason to look for 'cheerful confidence in one's own strength and ability' in Charles's work. This in itself is a sufficient reason why his writings could never have achieved the same popularity as those of Jomini or Clausewitz. The centre of military thought has normally tended to follow the centre of military power. Thus, during France's period of greatness, it was Jomini's interpretation of Napoleon's bold strategy that was studied throughout the Western world. And when Germany became the major power in Europe with a supreme military orientation, a German military school presenting Clausewitz as its forerunner dominated military thought and the interpretation of military history.

II JOMINI: SYNTHESIZING THE LEGACY OF THE ENLIGHTENMENT WITH NAPOLEONIC WARFARE

Background and Early Development

Antoine Henri Jomini (1779–1869) synthesized the theoretical ideal of the Enlightenment with Napoleonic warfare, producing a penetrating and fertile rationale of the new type of operations. Hence both the enormous success and influence of his work in the nineteenth century and its decline in the twentieth.

Unfortunately, since the work of the military thinkers of the eighteenth century lost most of its relevancy in post-Napoleonic warfare and fell into oblivion, the origins of Jomini's theoretical outlook have become totally obscure. Though Jomini gave one of the most comprehensive descriptions of contemporary military literature in the introduction to his *Summary of the Art of War* and made several references to his predecessors in other works, modern readers have usually been unfamiliar with the thinkers cited and particularly with their characteristic theoretical outlook. Moreover by doing his utmost to emphasize his originality, Jomini did not make the work of historians easier. Consequently, one study called Jomini the 'Adam Smith' of military thought, and in attempting to trace his intellectual origins which clearly pointed to the Enlightenment, searched for his mentor among the most eminent philosophers of that age, and tended to find him in Montesquieu.[1] Another deemed to have discovered the source of his conception of theory in Lloyd.[2]

In fact, the intensely philosophical age of the Enlightenment was over, and Jomini, representing a new type of professional soldier, had little philosophical background or interest. He simply continued to follow the theoretical vision and conceptual framework established by all the military thinkers of the Enlightenment. He claimed no originality for his conception of theory but merely argued that he had finally filled an old theoretical ideal with real content.

[1] C. Brinton, G. Craig, and F. Gilbert, 'Jomini', in Earle (ed.), *Makers of Modern Strategy*, 79–80, 91.

[2] John Shy, 'Jomini', in Paret (ed.), *Makers of Modern Strategy*, esp. pp. 148–9.

Jomini was born in 1779 to a respected family in the town of Payerne, Canton Vaud, in French-speaking Switzerland.[3] Excited by the political and military events of the period, he left a banking and commercial career in Paris, and in 1798, using personal connections, he became secretary to the Minister of War of the Helvetic Republic, the French satellite, and rose to the rank of *chef de bataillon*. In 1801, he returned to his commercial and financial occupations in Paris, but his attention was now totally drawn to the military field.

In 1802 Jomini began reading extensively the works of the major military thinkers of the French Enlightenment 'commencing with Puységur, finishing with . . . Guibert'.[4] A year later he wrote his first work, composed of a set of maxims. He managed to bring it to the attention of Marshal Ney, who was impressed by the young Swiss and invited him to join his staff at the camp of Boulogne, though with no official appointment. At that time Jomini discovered the French translation of Lloyd, and Bülow's 'system', and decided to put his early work in the fire and write a new one in its place. This was the celebrated *Treatise* on the campaigns of Frederick the Great and the wars of the Revolution. The first two volumes of the *Treatise* were published in 1804–5 and three others followed until the completion of the first edition in 1809–10.[5]

As an unofficial member of Ney's staff, Jomini took part in the campaign of 1805. After Austerlitz and with Ney's assistance, he succeeded in passing the first volumes of the *Treatise* to Napoleon. The emperor's favourable impression of the work brought Jomini

[3] No scientific biography of Jomini has yet been written. From the three existing biographies, the earliest, written by Jomini's disciple and friend, the Swiss Colonel Lecomte, during Jomini's lifetime and, apparently, from his own mouth, is naturally, uncritical. However it constitutes the almost sole source of all subsequent biographies: Ferdinand Lecomte, *Le Général Jomini, sa vie et ses écrits* (Paris and Lausanne, 1860). The biography written by the celebrated French literary critic C. Sainte-Beuve, *Le Général Jomini, étude* (Paris, 1869), adds significant documents from the correspondence of Napoleon and Berthier. The latest biography, Xavier de Courville's *Jomini ou le devin de Napoléon* (Paris, 1935), written by a descendant of Jomini, adds virtually nothing to its two predecessors. On Jomini's family and youth see Jean-Pierre Chuad, 'Les Années d'enfance et de jeunesse d'Antoine Henri Jomini' in *Le Général Antoine Henri Jomini* (Lausanne, 1969), 11–24.

[4] Jomini, *Summary*, pp. 11–12; all references are made to the American trans. of the *Précis: Summary of the Art of War* (Philadelphia, 1862); However, references to the bibliographical introd. which is not included in this edn., are made to a previous trans. (New York, 1854).

[5] Ibid. 12–13; see n. 15 below for the *Traité*'s development and changes of title.

an official appointment in Ney's staff and a rank of colonel.[6] With Ney's corps he participated in the campaigns of 1806–7 in Prussia and Poland, rising after the Battle of Friedland to the position of chief of staff. In this capacity he took part in the campaign of 1808 in Spain. In December 1810, he was appointed brigadier-general, and in the Russian campaign of 1812–13 he became military governor first of Vilna, and later of Smolensk. The campaign of 1813 in Germany saw him back in his old position as chief of staff to Ney, with whom he fought the Battle of Bautzen. However, the rejection of Ney's recommendation to appoint him major-general, the famous hostility of Berthier, Napoleon's chief of staff, and alleged unjust treatment from the Emperor himself brought him to desert to the Allies, a course taken by several of Napoleon's generals during his years of decline. Jomini had already been offered an appointment in the Russian army in 1810, and now, in 1813, he became attached to Tsar Alexander's headquarters.

After the war, as a Russian general, Jomini's major works included the fifteen volumes of the *Histoire critique et militaire des guerres de la révolution* (Paris, 1820–4), the four volumes of the *Vie politique et militaire de Napoléon* (Paris, 1827) and the successive versions of his *Summary of the Art of War*, which marked the pinnacle of his growing reputation as the most important military theoretician of the era.

The early stages in the development of Jomini's ideas have remained largely obscure. Their study involves certain difficulties since the sole evidence is Jomini's own account which is highly tendentious and at times contradictory.

As mentioned, in 1802 Jomini began an extensive perusal of the works of the military thinkers of the French Enlightenment, 'commencing with Puységur, finishing with Mesnil-Durand and Guibert'. He found 'everywhere but systems more or less complete of the tactics of battles, which could give but an imperfect idea of war, because they all contradicted each other in a deplorable manner'.[7] This inevitable dismissal of all previous theoretical work, intending to prepare the ground for presenting the author himself as the founder of military science, is familiar enough already. It does include, however, a new theme. The preoccupation of the military

[6] See below: 'Jomini and Napoleon'.

[7] *Summary*, p. 12; see also p. 9.

thinkers of the Enlightenment with battle formation and deployment seemed now—with the shift of interest to the conduct of operations—most unsatisfactory. For this very reason, Jomini's discovery of Lloyd and Bülow appears to have left a deep impression on him which even his later attempts to minimize could hardly conceal. Indeed, according to his own account, his discovery of Lloyd and Bülow led to his decision to burn his early work and write another one in its place.

It is true that Jomini tried to explain this decision mainly on literary and technical grounds, confining his debt to Lloyd and Bülow merely to the inspiration of substituting a more vivid historical demonstration of his principles for an abstract presentation.[8] The forced nature of this argument is particularly patent in relation to Bülow's abstract work. Attempting to fix the consolidation of his ideas as early as possible and thus magnify his claims to originality, Jomini even wrote in the introduction to the third edition of the *Treatise* (1818) that the writing of this book had begun in 1802, ignoring the early attempt of 1802–3 of which we know, as mentioned above, only from his later and more confident account in the *Summary*.[9]

Throughout his life, Jomini never passed up an opportunity to belittle the stature of other military authors. When writing the *Treatise* in his twenties, struggling to consolidate his own identity against that of his predecessors, he was highly sardonic towards Lloyd and Bülow. Thirty years later, in 1837, when he was the most celebrated military theoretician of the period, he permitted himself, in the introduction to the *Summary*, little more generosity. Lloyd, he wrote,

> raised in his interesting memoirs important questions of strategy which he unfortunately left buried in a labyrinth of minute details on the tactics of formation and upon the philosophy of war. But . . . it is necessary to render him the justice to say that he first pointed out the good route. However, his narrative of the Seven Years War . . . was more instructive (for me at least) than all he had written dogmatically.[10]

Jomini's diplomatic treatment of Archduke Charles's work appears, however, to betray a more balanced and genuine appreciation of the significance of Lloyd's and Bülow's writings and of their

[8] Ibid. 12–13.
[9] Jomini, *Treatise on Grand Military Operations* (New York, 1865), p. xxi.
[10] *Summary*, pp. 10–11.

influence on the development of his own ideas. Charles's work, he wrote, 'put the complement to the basis of the strategic science, of which Lloyd and Bülow had first raised the veil, and of which I had indicated the first principles in 1805 in a chapter upon lines of operations'.[11] The concept of *lines of operations* Jomini did not invent; he inherited it. Lloyd's and Bülow's treatment of the conduct of operations gave his own theoretical work a decisive turn.

If a reconstruction of the early stages of Jomini's development is therefore attempted, it may be assumed that the early, abstract work from 1802–3, which was written after the study of the military thinkers of the French Enlightenment and which impressed Ney, was built around a set of maxims that, like the military works of the Enlightenment, were intended to cover all aspects of the treated field, which was now, apparently, the conduct of operations. The traces of this early theoretical stage can perhaps be identified in the many minor maxims scattered throughout the *Treatise* of 1804–5 alongside the major principles of operations, but diminishing in significance with Jomini's later development. These maxims deal, for example, with the defence of villages and their incorporation into the line of battle, the employment of cavalry along the edge of woods and on rugged terrain, the conduct of a besieging army which comes under attack, the securing of heights, and so on.[12]

However, the core of Jomini's early work seems to have been an embryonic formulation of a principle for the conduct of operations, applicable to both levels of battle and campaign. Describing the beginning of his intellectual development he wrote:

> Already had the narrative of Frederick the Great commenced to initiate me in the secret which had caused him to gain the miraculous victory of Leuthen. I perceived that this secret consisted in the very simple manœuvre of carrying the bulk of his forces upon a single wing of the hostile army; and Lloyd soon came to fortify me in this conviction. I found again, afterwards, the same cause in the first successes of Napoleon in Italy, which gave me the idea that by applying, through strategy, to the whole chess table of war, this principle which Frederick had applied to battle, we should have the key to all the science of war.[13]

Jomini's retrospective account is confirmed also by his earliest-known work, the *Treatise* of 1804–5: 'the conduct of the king at Leuthen',

[11] *Summary*, p. 13.
[12] *Treatise*, pp. 158, 162, 165–6, 422–3.
[13] *Summary*, p. 12.

he wrote, 'includes, in our opinion, the principle of all combinations in war'.[14]

With this principle Jomini gave theoretical expression to the new military reality and ideal raised to prominence by Napoleonic warfare: concentration of force. However, only the discovery of Lloyd's line and rationale of operations, turned by Bülow into the foundation of a new, albeit fantastic, science of operations, paved the way for Jomini's mature theoretical work. The fusion of the principle of the concentration of force with the rationale of operations gave him the key to the analysis of the Napoleonic art of operations, and this was precisely the synthesis achieved in the *Treatise* that Jomini began to write and publish in 1804–5.

The Conception of Theory and the Principles of Operations

After the period of consolidation (1802–4), Jomini's work reveals a remarkable continuity and consistency. The young man had completed most of his intellectual development by his late twenties, between 1804–9 when he wrote the *Treatise*, or in fact, by 1806–7 after experiencing the peak of Napoleonic warfare in the campaigns of 1805–7, and writing the summary of his principles which later became the concluding chapter of the *Treatise*.[15] Little was added

[14] *Treatise*, ch. VII, p. 252; see also p. 255.

[15] The confusion surrounding the exact development of the *Traité*, complicated by changing titles and differing eds. which Lecomte had failed to put in order (Lecomte, *Jomini*, pp. 321–2), was clarified by J. I. Alger's *Antoine Henri Jomini: A Bibliographical Survey* (West Point, 1975). The first two vols. on the first campaigns of the Seven Years War appeared in 1804–5 (see Alger, *Survey*, p. 2 for the divergence in dates) under the following title: *Traité de grand tactique, ou relation de la guerre de sept ans, extraite de Tempelhoff commentée et comparée aux opérations des dernières guerres, avec un recueil des maximes les plus importantes de l'art militaire*. The first vol. promised a seven vol. work: five vols. on the Seven Years War, a sixth on the campaigns of 1792–1800, and a seventh, a theoretical one. However, this plan was abandoned. On the publisher's request, the vol. on the first wars of the Revolution, to be the fifth vol. of the second edn., was next to appear: *Traité de grand tactique, relation critique des campagnes des Français contre des coalisés* (1806). The third and fourth vols., on the last campaigns of the Seven Years War, appeared in 1807 and 1809 respectively, under the new main title *Traité des grandes opérations militaires*. The article that summarized the principles, and was to become the concluding chapter of the *Traité*, was written in December 1806 when Jomini was stationed at Glogau in Silesia, and was published separately in 1807 as well as in Ruhle von Lilienstern's journal, *Pallas* (1808). The second edn. of the *Traité des grandes opérations militaires* included three additional vols. on the campaigns of the Revolution. The first six vols. of this edn. appeared in 1811, and the last two, delayed by the censure, were only published in 1816. From the third edn. of 1818, the campaigns of the Revolution were transferred to the *Histoire critique et militaire des guerres de la Révolution*.

or changed between the ideas of the twenty-eight-year-old colonel and those of the fifty-eight-year-old general and celebrated military thinker who published his *Summary of the Art of War* in 1837.[16]

As mentioned above, Jomini's conception of military theory was ready-made for him:

> The fundamental principles upon which rest all good combinations of war have always existed . . . These principles are unchangeable; they are independent of the nature of the arms employed, of times and places . . . For thirty centuries there have lived generals who have been more or less happy in their application . . . the battles of Wagram, Pharsalia and Cannae were gained from the same original cause.[17]

> Genius has a great deal to do with success, since it presides over the application of recognized rules, and seizes, as it were, all the subtle shades of which their application is susceptible. But in any case, the man of genius does not act contrary to these rules.[18]

Unfortunately, according to Jomini, his predecessors in the search for a universal theory of war had erred in looking for complete systems, some to the point of ridicule, thus inviting scepticism regarding fixed principles.[19] The need therefore arose for a 'demonstration of immutable principles and . . . [for] establishing a common standard from opinions which had differed so widely. It has been my fortune', wrote Jomini, 'to undertake this difficult task.'[20]

The same theoretical outlook was reiterated in the *Summary* of 1837:

> There exists a small number of fundamental principles of war, and if they are found sometimes modified according to circumstances, they can nevertheless serve in general as a compass to the chief of an army . . . Natural genius will doubtless know how, by happy inspirations, to apply principles as well as the best studied theory could do it; but a simple theory . . . without giving absolute systems . . . will often supply genius and will even serve to extend its development.[21]

On one central point, however, Jomini's theoretical outlook fundamentally differed from that of his predecessors. This divergence

[16] The *Introduction à l'étude des grandes combinaisons de la stratégie et de la tactique* (Paris, 1829) constituted an early version of the *Summary*. A more developed version, *Tableau analytique des principales combinaisons de la guerre, et de leur rapports avec la politique des états*, appeared a year later.

[17] *Treatise*, the concluding ch., p. 445.

[18] Ibid., ch. 7, pp. 253–4.

[19] Ibid., the concluding ch., pp. 446–7.

[20] Ibid. 447.

[21] Ibid. 18.

was related to the shift in emphasis from tactics to strategy, but the explanation for it lies deeper. The military thinkers of the Enlightenment maintained that the details of military organization and deployment were fundamentally mechanical and could be fixed in a definitive system, whereas the conduct of operations was almost entirely in flux, and therefore fell chiefly in the realm of the general's ingenious inspiration. This conception can still be found in the *Summary*, for instance, in the statement that 'the most minute . . . the most accessory points of tactics [are] the only part of war, perhaps, which it is possible to subject to fixed rules'.[22] However, a new approach clearly dominated this work, an approach which was to characterize military thought in the nineteenth and twentieth centuries and in which strategy could be reduced to universal principles while tactics were difficult to regulate and were exposed to constant transformations.

The interplay of moral forces which could hardly be foreseen was described by Jomini as one factor that prevented a general theoretical determination of tactics.[23] The main reason for the reversal of opinions was, however, the increasing awareness by the men of the period of the significance of techno-tactical developments that had been virtually ignored by the military thinkers of the Enlightenment with their universal frame of mind. During his career, Jomini witnessed changes in tactics which made battle formation more flexible and which rendered the military theory of the eighteenth century clearly outdated. Tactics could no longer be reduced to rigid patterns, he wrote in the *Summary*.[24] Napoleon had the same idea in mind when he remarked that tactics had to change every ten years. By the end of his long life, Jomini saw revolutionary technological innovations whose potential effect on the field of tactics was far-reaching. Some of the developments in armament, such as the Congreve rocket, the howitzer firing shrapnel shell, Perkins's steam-gun, and the improved musket, had already attracted his attention in the *Summary* of 1837.[25] Only strategy appeared to escape change:

The new inventions of the last twenty years seem to threaten a great revolution in army organization, armament and tactics. Strategy alone will remain unaltered, with its principles the same as under the Scipios and Caesars,

[22] Ibid. 9. [23] Ibid. 321.
[24] Ibid. 195. [25] Ibid. 48, 299.

Frederick and Napoleon, since they are independent of the nature of the arms and the organization of the troops.[26]

Tactics can still be studied theoretically by rules and principles; but 'strategy particularly may be regulated by fixed laws resembling those of the positive sciences'.[27]

The growing impact of the industrial revolution did not however stop with tactics. At the end of his life, faced with new challenges to the whole of his theoretical outlook, Jomini argued that the growing military use of railways could not change his universal principles of strategy.[28]

There was therefore some justification for Jomini's claim to be the founder of 'real' military science. The conduct of operations, he maintained, had to be the focus of the theoretical effort.[29] The attempt to reveal its principles was his central aim, and again shows remarkable continuity from the first volumes of the *Treatise* in 1805–7 to the *Summary* of 1837.

The *Traité de grand tactique* or, in its later title *Traité des grande opérations militaires*, was based on Lloyd's and Tempelhoff's accounts of the Seven Years War, and took issue with their contending interpretations. The core of the work was, however, a set of principles for the conduct of operations, which had at first been derived from some dominant features of Frederickian warfare, but which were soon adapted to reflect the Napoleonic rationale of operations. These principles were scattered throughout the *Treatise* but were also concentrated in several summarizing chapters, whose differing dates of publication help to trace the final stages in the development of Jomini's ideas. In the first two volumes of the

[26] *Treatise*, p. 48, see also p. 9. Archduke Charles developed a similar conception at roughly the same time. In contrast to the military thinkers of the French Enlightenment, he too wrote that 'strategy is the science of war' whereas 'tactics is the art of war' and applies the principles of strategy to changing circumstances and new inventions: Archduke Charles, *Grundsätze der Strategie*, *AS* i. 235; *Grundsätze der höheren Kriegskunst*, *AS* i. 50.

[27] *Summary*, p. 321.

[28] Lecomte, *Jomini* (3rd edn. 1894); cited by Alger, *Survey*, p. 19.

[29] Jomini fully adopted the concept of 'strategy' and the distinction between 'strategy' and 'tactics' along the lines introduced by Bülow only in the *Summary*, after they had been accepted in German military literature and used both by Archduke Charles and Clausewitz. When writing the *Treatise*, he still employed the traditional concepts of 'grand tactics' or 'operations' for the whole conduct of operations. He also lent his principles equal validity in the conduct of both battle and campaign.

Treatise, published in 1804–5, chapter 7 provided an initial summary of the principles of operations, further developed in chapter 14, 'Upon Lines of Operations'. These were the chapters that impressed Napoleon after Austerlitz.[30] In December 1806, having experienced Napoleon's great campaigns, Jomini reformulated his principles with minor additions in an essay which was to become the concluding chapter of the *Treatise*.

The essence of Jomini's work lay in revising Lloyd's rationale of operations in terms of the new Napoleonic warfare, and in reformulating Bülow's vision of strategic science into a more moderate and common-sensical form. We have seen that Bülow had already had to contend with the collapse of the traditional conduct of operations when confronted by Napoleonic strategy. Jomini, who was not tied by past prestige to old conceptions, could make these adjustments without any inhibitions.

Firstly, the premises of the rationale of operations itself had to be revised. The lines of operations, wrote Jomini, 'have been considered merely in their material relations. Lloyd and Bülow have only attached to them the importance which pertains to the magazines and depots of an army.'[31] This conception had, obviously, been exaggerated even in relation to eighteenth-century warfare; and now, with the heavy reliance on the countryside and the new emphasis on decision, it became totally inadequate. This, however, did not imply that the lines of operations lost their importance. 'The greatest secret of war consists in becoming master of the communications of the enemy'; Jomini attributed this statement to Napoleon himself.[32] Only the function of these lines changed with the new type of warfare. Now, they were not perceived merely as lines of supply, but also, or perhaps chiefly, as lines of retreat and communications with friendly forces. The seizure of the enemy's communications could lead not only to his starvation and withdrawal but also to his destruction.

The destruction of the enemy's field armies was the new military aim, necessitating a second adjustment in Lloyd's and Bülow's rationale of operations. In 1804–5 when the two first volumes of the *Treatise* were published, Jomini's position was still relatively conservative: one has 'to give battle only when great advantages are to be derived, or the position of the army makes

[30] See below 'Jomini and Napoleon'.
[31] *Treatise*, ch. 14, p. 12.
[32] Ibid., ch. 15, p. 59.

it necessary'.[33] However, after the great decisive campaigns of 1805–7 his position became bolder. 'The art of war', he wrote, rejecting the positions of Lloyd and Bülow, 'does not consist in running races upon the communications of our enemy, but in the securing of them, and marching thereon, for the purpose of bringing him to battle.[34] Furthermore, 'after the victory, the vanquished should be allowed no time to rally, but should be pursued without relaxation'.[35] For all that, Jomini never considered battle to be the almost exclusive means of war as Clausewitz did. 'Battles have been stated by some writers to be the chief and deciding features of war,' he wrote in the *Summary* in direct reaction to Clausewitz's ideas; 'This assertion is not strictly true, as armies have been destroyed by strategic operations without the occurrence of pitched battles.'[36]

Finally, the rest of the new military values were stressed. Initiative was placed by Jomini at the head of his principles in 1804–5 and again in 1806–7.[37] Mobility and movement were other fundamental features of Napoleonic strategy: 'the system of rapid and continuous marches multiplies the effect of an army'; indeed, a 'march of thirty miles a day' was presented by Napoleon in his famous dictum as one of the three components of his art of operations.[38] Most important of all was the concentration of force, Jomini's earliest principle. 'The employment of masses upon the decisive points', he wrote, 'constitutes alone good combinations, and . . . it should be independent of all positions'.[39]

These, then, were the changes and revisions that the body and spirit of the old rationale of operations had to undergo in order to be accommodated to the Napoleonic form of operations. This was the basis for Jomini's new synthesis. The secret of operations, in the conduct of both battle and campaign, lay in the concentration of maximum force to achieve local superiority at the decisive point. Initiative and forced marches were means to this end;[40] but the real secret of the operational formula lay in the skilful use of the lines of operations:

[33] *Treatise*, ch. 13, p. 443.
[34] Ibid., ch. 27, p. 323.
[35] Ibid., ch. 13, p. 443.
[36] *Summary*, p. 178. For Clausewitz's views on this matter see Ch. 7 of this book.
[37] *Treatise*, ch. 5, principle 1, p. 201; the concluding ch. principle 1, pp. 448–9.
[38] *Summary*, pp. 176, 137.
[39] *Treatise*, ch. 3, p. 149; see also ibid., ch. 7, p. 252, and *Summary*, 'The Fundamental Principles of War', p. 71.
[40] *Treatise*, ch. 7, p. 252; *Summary*, pp. 72–3, and p. 176, principle 6.

If the art of war consists in bringing into action upon the decisive point of the theatre of operations the greatest possible force, the choice of the line of operations (as the primary means of attaining this end) may be regarded as fundamental in devising a good plan for a campaign.[41]

The choice of the most advantageous line of operations was therefore the centre of Jomini's work, and already in 1804–5 he had put forward two main patterns to which he remained loyal throughout his life:

It would generally be better to direct the line of operations upon one extremity, whence we can at will reach the rear of the enemy's line of defence. The direction upon the centre is best only when the adversary's line is very long, and the different corps which guard it separated by long intervals.[42]

Thus, when two armies confront each other, both on the battlefield and in the theatre of operations, operations should usually be directed against one of the extremities of the enemy's front and towards his communications with his rear. Jomini's initial idea, patterned on the model of Leuthen, may have been focused on the battlefield and the destruction of the enemy's wing.[43] However, with his discovery of Lloyd and Bülow, and Napoleon's Marengo campaign of 1800, he seems to have switched the emphasis to the strategic scale and the manoeuvre around the enemy's flank, in an attempt to seize his communications, cut him off, and destroy him. 'The combinations of the campaign of 1800', he wrote in 1804–5, 'have clearly demonstrated the truth of this maxim.'[44] And Napoleon's great campaigns of 1805–7 strikingly reinforced it: the manœuvre towards the enemy's rear and against his communications—*la manœuvre sur les derrières* in Camon's classic typology, influenced by Jomini's conceptions—led to the capitulation of Mack's army at Ulm (1805) and to the destruction of the Prussian army at Jena-Auerstädt (1806). Both in 1800 and 1806 the fate of the war was decided in a single, all-embracing blow. Following Jena, Jomini began to stress the advantages of manœuvring the enemy against impassable obstacles; attacked from his flank and rear, he faced total destruction.[45]

[41] *Summary*, p. 113; and p. 176, principle 6.

[42] *Treatise*, ch. 5, principle 3, pp. 201–2; see also ch. 14, principle 4, pp. 42–3, and in the concluding ch., principle 2, pp. 449–50.

[43] *Treatise*, ch. 7, p. 252, and Jomini's later account in the *Summary*, p. 12.

[44] *Treatise*, ch. 14, p. 43; and *Summary*, p. 12.

[45] For the example of 1806 and its abstract expression, see mainly the *Summary*, p. 115.

The manœuvre against the enemy's rear was one of the most impressive, and certainly the most decisive, forms of Napoleonic strategy.

Yet, when the enemy's front, both on the battlefield and in the theatre of operations, was over-extended or even composed of several separate corps, a single, great outflanking movement was ruled out. In this case another course of action became available which, although it did not threaten the enemy's communications and consequently was not as destructive as manœuvring against the enemy's rear, still used the particular pattern of these communications to achieve a decisive superiority at the point of engagement. The attacker could break through the enemy's front, or, if the enemy was deployed in several separate corps, penetrate between them. He was, then, concentrated (that is, he 'used a single line of operations') in a 'central position', and operated on 'interior lines' against a divided enemy operating on 'several exterior lines'. Each of the enemy's corps could then be defeated separately.

That was Frederick's natural position in the Seven Years War and the secret of his celebrated manœuvres which aimed to crush each of the armies of the coalition in succession. That was also the position in which Napoleon strove to place himself—with brilliant success—in his first campaign of 1796, when he penetrated between the armies of Piedmont and Austria in the Ligurian Alps, beating the former and driving her out of the war, and then turning against and defeating the latter. He again operated in interior lines when he marched to crush the Austrian armies which descended separately from the passes of the Alps to raise the siege from the fortress of Mantua (1796–7).

Jomini developed his conceptions of the central position and interior lines as one of the most important lessons of his study of the Seven Years War, which was confirmed by Napoleon's strategy. He thus regarded these forms of warfare as having a decisive advantage, and his great fame became chiefly associated with them. 'An army whose lines are interior and nearer than those of an enemy,' he wrote in 1804–5, 'can by a strategic movement, overwhelm those of the enemy successively . . . It follows from this that a double line of operations . . . would always be dangerous and fatal,' whereas 'single interior lines of operations are always the most sure.'[46] These were also the conclusions of Jomini's famous summary of his principles in 1806–7.

[46] *Treatise*, ch. 7, principles 2, 4, 6, pp. 249–50; See also ch. 14, principles 2 and 3, p. 42; and p. 39.

Jomini restated his double conception of operations in the *Summary*:

It may be laid down as a general principle that the decisive points of manœuvre are on that flank of the enemy upon which, if his opponent operates, he can more easily cut him off from his base and supporting forces without being exposed to the same danger . . . If the enemy's forces are in detachments or are too much extended, the decisive point is his centre, for by piercing that, his forces will be more divided, their weakness increased, and the fractions crushed separately.[47]

The former course of action 'led to the success of Napoleon in 1800, 1805 and 1806; the latter was successful in 1796, 1809 and 1814'.[48] Jomini's conception of operations was in essence, then, a formal presentation of the Napoleonic art of war at its heyday; that was the source of its power but also of its limitations.

Challenges and Criticism

Jomini claimed to have revealed the principles of Napoleonic warfare which were at the same time *also* the universal principles of the art of war. This double status was based on the belief that Napoleon's genius actually embodied the universal principles of war. There was a latent tension here which was only revealed when the Napoleonic system of war was shown to be less than perfect. As with the categories of neo-classicism, the universal validity of Jomini's principles, particularly his doctrine of the central position and interior lines, was thrown into question when significant exceptions could be presented against them. These were provided by Napoleon's last campaigns in the years 1813–15, and particularly by the great autumn campaign of 1813.

In 1814 Napoleon operated against the Allied armies that invaded France, with a virtuosity and speed displayed in his best campaigns. He attacked them successively, conducting forced marches between Schwarzenberg's Austrian army in the east, Blücher's Prusso-Russian army in the north-east, and the Prussian, Russian, and Swedish forces of the Army of the North. This, however, did not save him from defeat which was indeed almost unavoidable in view of the absolute numerical superiority of the Allies.

[47] *Summary*, p. 88; see also pp. 90, 114, 175–6.
[48] Ibid., 'Summary of the Principles of Strategy', p. 6.

In 1815 Napoleon attempted his classic penetration strategy, breaking through the centre of the Allies' line between the British and the Prussians in order to push them back in opposite directions and crush them separately. He almost succeeded in this when he defeated the Prussians at Ligny and turned to crush the British at Waterloo. The poor co-ordination of the French army was partly responsible for Napoleon's failure, but the counter-strategy of the Allies was no less influential. Unlike their conduct in Napoleon's early campaigns, the divided enemy armies, after their initial surprise, did not leave the initiative to the French. The Allies' attempt to achieve a forward concentration of forces had failed, but after the defeat at Ligny, the Prussians avoided the expected north-eastern retreat which would have moved them away from the British. In a determined action, Blücher succeeded in reuniting with Wellington on the battlefield of Waterloo, destroying Napoleon's campaign plan.

In 1814, although Napoleon operated with a clearly inferior force, he nevertheless achieved impressive successes; and in 1815 his subordinates, Grouchy and Ney, could be blamed for the failure. This, however, could not be said about the gigantic autumn campaign of 1813 in Germany. The French army and the armies of the Coalition were almost equal in size, but Napoleon, who kept his forces united, failed to crush his enemies separately. The powers of Europe, who for seventeen years had experienced the destructiveness of the Napoleonic art of war, prepared their homework very carefully. As an antidote to Napoleon's strategy of interior lines they refused to be exposed to his main thrust, and took counter-initiatives. The army against which Napoleon concentrated his forces, refused battle, withdrew, and drew the enemy after it. Napoleon's blow, therefore, struck thin air. Simultaneously, the rest of the Allied armies moved against the French rear, exerting superior pressure on the delaying corps left by Napoleon to screen his movements, and forcing his main thrust to halt and withdraw in order to confront the threat. Not only did Napoleon exhaust his forces in vain attempts to defeat the Allied armies separately, but he also failed to prevent their concentration on the battlefield at Leipzig. Furthermore, the new technique highlighted some inherent advantages of the exterior lines and a certain paradox in Jomini's thought. The side operating in exterior lines surrounded his opponent from several directions. He was, therefore, in a much better position to envelop his opponent and threaten his communications. The campaign of 1813 thus aroused extensive debate regarding the validity of one of Jomini's principal ideas.

Jomini was forced on the defensive. While he did not retract the statements attributing the advantage to interior lines, he did reformulate them more cautiously and claimed that he had never rejected operations on exterior lines under certain circumstances.[49] He argued that exceptional conditions prevailed in Napoleon's last campaigns: that Napoleon was numerically inferior, and that both sides used huge armies which, on the one hand, made it difficult for Napoleon to achieve rapid concentration of force, and on the other, increased the ability of each Allied army to resist independently. In principle, he argued that a few exceptions were not sufficient to invalidate a rule based on the main body of military experience.[50] What then should be regarded as the main body of military experience and what are unusual circumstances? What was the rule and what were the exceptions?

Once again the challenge of a new historical experience, rather than theoretical reasoning, undermined Jomini's conception of the central position and interior lines. In the great German wars of unification, both against Austria and against France, the Prussian army deployed several armies by rail in exterior and enveloping lines of operations and with decisive results. New means of communication, such as the telegraph, facilitated the co-ordination of the separate armies. A new controversy therefore broke out in Germany as to whether the Moltkean strategy, the new military model, introduced a new rationale of operations, completely different from the Napoleonic one and based on the superiority of exterior lines.[51] Which part of the historical experience was the 'correct one'?

What then was the problem with Jomini's theoretical work? The problem was that he regarded his conceptions, which were a penetrating schematization of the Napoleonic form of operations, to be a universal military theory. Jomini's great achievement was that he provided his contemporaries, who were striving to grasp the nature of the new type of warfare, with the clearest, most instructive conceptual framework for this task. His conceptions have since stood

[49] *Summary*, pp. 113, 126–7. [50] Ibid. 123–8.

[51] For a summary of the controversy, see Caemmerer, *Strategical Science*, pref. and ch. VIII.

behind every major interpretation of the Napoleonic art of operations, and proved highly valuable for soldiers as long as the major features of this form of warfare continued to prevail.[52] However, as a highly successful reflection of a particular period, Jomini's conceptions were far from being universally valid. Despite Bülow's boundless theoretical claims, he regarded his system as a product of the particular conditions of the modern period—the reliance on a system of supply. Conversely, Jomini, a true child of the Enlightenment, limited his principles by leaving room for creative application, circumstances, chance, and the like, but regarded them as valid in every time and place.

The result was an encapsulation of all the problems for which the famous 'unhistorical approach' of the Enlightenment was blamed.[53] Jomini claimed that all military history from 'Scipio and Caesar to Napoleon' had been guided by the principles that he extracted from Napoleonic warfare, and referred to all periods of history that clearly contradicted this claim as undeveloped or degenerate. For instance, he did not perceive the complicated supply system of the eighteenth century as the product of the particular conditions of that period, but rather viewed it as an error caused by inadequate thinking and 'prejudice' that even Frederick the Great was unable to shake off.[54] The same line of reasoning applied to strategy. Rather than understanding Frederick's strategy against the background of the political and military conditions of his time, Jomini maintained that Frederick had not operated according to Napoleonic principles because military thought had not yet developed enough to recognize these principles. 'Until Frederick's time but little was known except concerning' tactics, and 'the fact is, that the art of war made but little progress' even under Frederick. Frederick 'entirely misunderstood' the principles of operations.[55]

The vast majority of Jomini's contemporaries shared the fundamentals of his theoretical outlook as established by the military thinkers of the Enlightenment. Criticism of his ideas, particularly

[52] The two major interpretations of the Napoleonic art of war, that of Camon in France and York von Wartenburg in Germany, owed much to Jomini's categories of analysis. See York von Wartenburg, *Napoleon as a General* (London, 1902), i. 276–7, 299 for praises for Jomini.

[53] See Chs. 5.I and 6.II of this work.

[54] *Treatise*, ch. 1, p. 93; ch. 3, p. 164.

[55] Ibid., ch. 26, p. 278.

that of interior lines after the campaign of 1813, was limited to the details (albeit important ones) rather than the essence of his theoretical work. However, with Clausewitz who rejected the entire military tradition of the Enlightenment, the very legitimacy of Jomini's theoretical approach was denied. It was not that Clausewitz thought that 'Jomini said something which was utterly wrong'; indeed, compared with Bülow, 'it cannot be denied that he thinks and argues in an extremely more solid manner'.[56] Furthermore, until 1813, Clausewitz shared Jomini's belief in the primacy of the central position and interior lines.[57] However, Clausewitz was interested not so much in the certain practical value of Jomini's ideas but in their presentation as a general science of war which he regarded as absurd. Jomini's abstract principles, he argued, ignored the living reality of war, the operation of moral forces, and the unique conditions of every particular case. Jomini's criticism of Frederick the Great, for example, was totally unhistorical and superficial. The complicated options facing Frederick in highly complex situations could not be reduced to, or judged by, a couple of abstract principles.[58]

Clausewitz's early notes on strategy were not published until 1937, but his major works which appeared in 1832–7, expressed the same ideas and made them public. In *On War* he again argued that principles like those of Jomini which were abstracted from the real conditions of particular cases could never be universally valid, and that Jomini's criticism of other periods for not acting according to the Napoleonic system of warfare was strictly forbidden.[59] In a more direct reference to Jomini's work, Clausewitz's judgement was short and harsh:

> As a reaction to that fallacy [Bülow's] another geometrical principle was then exalted: that of the so-called interior lines. Even though this tenet rests on a solid ground—on the fact that the engagement is the only effective means in war—its purely geometrical character still makes it another lopsided principle that could never govern a real situation.[60]

That the conduct of war could not be reduced to universal principles was the general message of *On War*. War was affected by innumerable

[56] Clausewitz, 'Strategie', the important addition of 1808, in Hahlweg (ed.), *Verstreute kleine Schriften* pp. 48 and 47.

[57] See pp. 203 and 206 of this work.

[58] 'Strategie' (1808), pp. 47–9.

[59] Clausewitz, *On War*, VI, 30, p. 516.

[60] Ibid. II, 2, pp. 135–6.

factors, dominant among which were political conditions and moral forces; it was saturated with the unknown and incalculable, and was changing throughout history.

It is not hard to imagine that the publication of Clausewitz's works was an unpleasant surprise for Jomini. Completely unexpected, a huge and most impressive work, whose theoretical sophistication was indisputable, appeared on the scene, threatening Jomini's growing domination over military theory and challenging the accepted tenets of the study of war. It is apparent that Clausewitz's clear sense of superiority and the short, dismissive nature of his criticism wounded Jomini very deeply, even more than the general arguments of his work. Jomini's references to Clausewitz throughout the *Summary* imply as much. In his survey of military literature at the beginning of the *Summary*, Jomini made an effort to check his anger, and his treatment of Clausewitz was more elaborate and cautious than his customary treatment of any other military writer with the exception of Archduke Charles. However, his aim was to fight back and he neglected no opportunity or subtlety to belittle the value of his rival's work. In 1831, he wrote,

> the Prussian General Clausewitz died, leaving to his widow the care of publishing his posthumous works which were presented as unfinished sketches. This work made a great sensation in Germany . . . One cannot deny to General Clausewitz great learning and a facile pen; but this pen, at times a little vagrant, is above all too pretentious for a didactic discussion, the simplicity and clearness of which ought to be its first merit. Besides that, the author shows himself by far too skeptical in point of military science . . . As for myself, I own that I have been able to find in this learned labyrinth but a small number of luminous ideas and remarkable articles; and far from having shared the skepticism of the author, no work would have contributed more than his to make me feel the necessity and utility of good theories.[61]

Clausewitz's remarks about the many inaccuracies in Jomini's historical studies annoyed Jomini very much. He wrote bitterly that Clausewitz

> has been an unscrupulous plagiarist, pillaging his predecessors, copying their reflections, and saying evil afterwards of their works after having travestied them under other forms. Those who shall have read my campaign of 1799, published ten years before his, will not deny my assertions, for there is not one of my reflections which he has not repeated.[62]

[61] *Summary*, pp. 14–15. [62] Ibid. 21.

In addition, throughout the *Summary*, Jomini took issue with Clausewitz on a series of points, unleashing his wounded pride. Regarding mountain warfare he wrote about 'General Clausewitz, whose logic is frequently defective'. Concerning the importance of manœuvre in battle, he wrote that 'Clausewitz commits a grave error'.[63] Yet, Jomini's full response to the challenge of Clausewitz, as well as to that of Archduke Charles, is revealed mainly indirectly in the *Summary*.

Jomini's Response and Later Development

Clausewitz as a rival and Archduke Charles both as an ally and contender played an important role in Jomini's decision to write the *Summary*. Jomini himself wrote that the publication of Charles's *Grundsätze der Strategie* (1814) convinced him of the need to supplement his historical *Treatise* with an abstract theoretical work. This led to the first early version of the *Summary*, the *Introduction à l'étude des grandes combinaisons de la stratégie et la tactique* (1829).[64] When yet another impressive and comprehensive theoretical work such as *On War* was published, Jomini could hardly fall behind. His aim in writing the *Summary* was, therefore, to produce a comprehensive theoretical work of his own which would include and surpass what he regarded as Charles's main achievements, and repel the challenge posed by Clausewitz.

Although the *Treatise* was mostly historical and full of references to many aspects of war, it was dominated by Jomini's principles of operations, the source of his great reputation as the interpreter of the Napoleonic art of war. As the *Summary* was intended to involve a more comprehensive and rounded treatment of war, it was more heterogeneous in nature. The principal theoretical additions were the first two chapters, 'The Relation of Diplomacy to War' and 'Military Policy'. The former 'included those considerations from which a statesman concludes whether a war is proper, opportune, or indispensable, and determines the various operations necessary to attain the object of the war'.[65] A typology of aims in war follows (rights, economic interests, balance of power, ideology, territories, and mania for conquest), together with a classification of the various types of war. The second chapter, 'Military Policy',

[63] Ibid. 166, 178. [64] Ibid. 14. [65] Ibid.

embraces the political considerations relating to the operation of armies . . . the passions of the people to be fought, their military system, their immediate means and reserves, their financial resources, the attachment they bear to their government or their institutions . . . finally, the resources and obstacles of every kind likely to be met.[66]

Contrary to what might be assumed, these chapters were not written under Clausewitz's influence. They had appeared for the first time in the second early version of the *Summary*, the *Tableau analytique des principales combinaisons de la guerre, et de leur rapports avec la politique des états* (1830), written before the publication of *On War*. In fact, the theoretical framework developed in 1830 had already been outlined in the third edition of the *Treatise* in 1818. At the end of the concluding chapter of principles Jomini wrote:

It is not necessary to remind our readers that we have here merely treated of those principles which relate . . . to the purely military part of the art of war; other combinations no less important . . . pertain more to the government of empires than the commanding of armies. To succeed in great enterprises it is . . . necessary . . . to take into consideration the resources . . . internal condition . . . the relative situation of their neighbors . . . the passions of the people . . . their peculiar institutions and the strength of their attachment to them . . . In a word, it is absolutely necessary to know that science which consists of a mixture of politics, administration and war, the basis of which has been so well laid by Montesquieu.[67]

Like Guibert and Lloyd, and even perhaps under their influence, Jomini applied the theoretical legacy of Montesquieu to the military field.

Some additions were also built around the rationale of operations itself, mainly under the influence of Archduke Charles. Jomini's principles of operations in the *Treatise* were renowned for being remarkably simple, clear, and concise in formulation. Yet, because of his effort to match and even surpass Charles's system of strategic points, they became in the *Summary* much more elaborate, less sharp in presentation, and entangled in complex definitions and jargon. Jomini had tended always to identify definitions with the scientific approach. In the *Treatise* he had already defined single, double, interior, exterior, extended, deep, concentric, eccentric, secondary, and accidental lines.[68] Charles's work only reinforced this tendency

[66] *Summary*, p. 38. [67] *Treatise*, the concluding ch., pp. 460–1.
[68] *Treatise*, ch. 14, pp 11–12.

in Jomini. He was not prepared to fall behind Charles in what he regarded to be one of the fundamentals of the science of war—a systematic approach in characterizing the theatre of operations. In the *Summary* he therefore identified and defined fixed and accidental-intermediate bases of operations; objective, strategic-manœuvre, decisive-strategic, geographic-strategic, and refuge points; operational and strategic fronts, and strategic positions, zones, and lines of operations; and temporary, strategic, and communication lines.[69] All this jargon originated late, and, as noted by a modern commentator, only obscured Jomini's central teaching.[70] Moreover, it strengthened the criticism that his work was pedantic and geometric.

Jomini responded to the criticism that his approach was mechanistic and geometric with somewhat justified astonishment and bitterness. He believed that he was accused of opinions he had never held:

> My principles have been badly comprehended by several writers . . . some have made the most erroneous application of them . . . others have drawn from them exaggerated consequences which have never been able to enter my head; for a general officer, after having assisted in a dozen campaigns, ought to know that war is a great drama, in which a thousand physical or moral causes operate more or less powerfully, and which cannot be reduced to mathematical calculations . . . I hope that after these avowals, I could not be accused of wishing to make of this art a mechanism of determined wheel-work, nor of pretending, on the contrary, that the reading of a single chapter of principles is able to give, all at once, the talent of conducting an army.[71]

It is important to understand Jomini's reaction to the criticism against him. His theoretical outlook was based totally on the intellectual legacy of the military thinkers of the Enlightenment, accepted for a century as self-evident. In view of this, the criticism against him which expressed new intellectual trends was unintelligible to him. Military views, as such, were not so much the issue, but rather differing perspectives regarding the question as to which theoretical approach was at all legitimate and worth pursuing.

[69] *Summary*, pp. 74–132.

[70] Michael Howard, 'Jomini and the Classical Tradition in Military Thought', in Howard (ed.), *The Theory and Practice of War*, pp. 16–17.

[71] *Summary*, pp. 17–18.

Like all the military thinkers of the Enlightenment, Jomini was always aware of the importance of moral and immeasurable factors but regarded them as belonging to the 'sublime' part of war which was not susceptible to scientific treatment and comprised the 'art' component in the conduct of war. It is true that Lloyd studied the psychological motives of troops with a practical purpose in mind, but on the whole the military thinkers of the Enlightenment saw no use in elaborating upon the moral, incalculable, and unforeseen which could hardly produce practical results.

The same applied to the criticism of the alleged geometrical nature of Jomini's work, expressed in particularly harsh words in Clausewitz's *On War*. Here too, Jomini's claim that he was unfairly treated, can be understood. Since the writing of the *Treatise*, Jomini always stressed that he did not believe in a military 'system', certainly not in a complete geometrical system such as that of Bülow. Instead, he believed in 'principles', or, in other words, in a much more flexible and less pretentious theoretical framework. This was not merely empty rhetoric or intellectual hair-splitting. Jomini's principles of operations were simple, relatively undogmatic, and faithfully expressed the spirit of contemporary warfare. His comment on the conceptions associated chiefly with Bülow and Charles was in this respect characteristic: 'They want war too methodical, too measured,' he wrote; 'I would make it brisk, bold, impetuous, perhaps sometimes even audacious . . . to reduce war to geometry would be to impose fetters on the genius of the greatest captains and to submit to the yoke of an exaggerated pedantry.'[72]

Indeed, Jomini never based his work on geometrical considerations in the full sense of the term. Unlike Bülow and Charles, he was far from forming his principles on calculations of angles and distances, and he repeatedly emphasized that the sketches in his works should be regarded as no more than illustrations.

To clarify this important point it might be worthwhile to distinguish between 'geometrical' and 'spatial' military theory. Since the shift of interest to the conduct of operations, the attention of military thinkers had been focused on the relationship between the movements of armies in space. As we have seen, this line of thought led to the formulation of several rationales of operations which were sometimes very fruitful. Lloyd schematized the rationale of the war

[72] *Summary*, p. 135.

of manœuvre, and Jomini developed an even more successful schematization of the Napoleonic art of operations. However, one had to be aware that the 'spatial' approach was but an abstraction and was therefore limited in validity. This was particularly important since the approach invited dangerous temptations. First and foremost, there was the tendency to give the spatial relationships a geometrical expression, thus realizing a quest that was deeply rooted in the Enlightenment, and making the rationale of operations precise and strictly scientific. This tendency found its most extreme manifestation with Bülow, but hardly touched Jomini. While he certainly expressed the universal tendency of the Enlightenment, Jomini was absolutely free from geometrical dogmatism.

If one may say so, Jomini's death in 1869 came at the most suitable time from the point of view of his international reputation. In old age, during the last decades of his long life, he was able to enjoy his position as the most celebrated and influential military thinker of his times. Archduke Charles had died twenty-two years earlier, and, in any case, his works were never so influential throughout Europe as those of Jomini. Clausewitz's works, whose sensational publication in Germany had alarmed Jomini in the 1830s, seemed to have fallen into respectful oblivion. The many eminent persons of the Napoleonic era with whom Jomini was on bad personal terms were also all dead. And new generations of officers, educated at military academies on Jomini's theoretical works and monumental histories, treated him with the admiration reserved for classic authors. Willisen and Rüstow in Germany, Napier and Hamley in Britain, and Mahan in the United States are only some of his most notable and declared disciples in the nineteenth century.[73]

Yet, immediately after Jomini's death, the picture began to change. Following the crushing victories of the Prussian armies under the orchestration of one who declared himself to be Clausewitz's disciple, the centre of military power and thought passed to Germany. A new, active, and powerful German military school influenced military thought throughout Europe, presenting Clausewitz as its mentor and projecting a particular interpretation of his ideas. Consequently,

[73] Jomini's wide-ranging influence in the 19th cent. still deserves to be studied; for an outline see Howard, 'Jomini', in Howard (ed.), *The Theory and Practice of War*, p. 14; and Shy, 'Jomini', in Paret (ed.), *Makers of Modern Strategy*, pp. 177–9.

Jomini's star was in eclipse. He gained a dubious image in Germany which absorbed and assimilated Clausewitz's criticism of universal doctrines of operations, criticism that was reinforced by Moltke's attitude. Jomini's reputation was also clearly declining in France which had previously been the military centre of Europe, and where, after the French defeat, the army was eager to learn the secrets of German military proficiency. In the peripheral countries, however, like the United States and Britain which were less sensitive and slower in reacting to the changes in the military centre of gravity, Jomini's dominance remained unchallenged. On the eve of the First World War in a course of lectures presented before the students of Oxford, Spenser Wilkinson, the University's first professor of military history, declared that the science of war had advanced very little since Jomini; Jomini had formulated the principles used in the study of military operations.[74]

The First World War was the second and more crucial turning-point in Jomini's decline, because in many ways it ended the Napoleonic model of warfare. In the Napoleonic-Jominian paradigm, still sufficiently relevant throughout the nineteenth century, armies manœuvred against one another in a relatively open space. In contrast, those who took part in the fighting on the Western front, when searching for an analogy in previous experience, could only describe it as a gigantic siege. The growing armies with their increasing fire-power filled space from one end to the other, blocking all movement with long and continuous front lines. The revival of the war of movement in the Second World War brought back some relevancy to the Jominian categories of manœuvre. But warfare was now conducted with mechanized armies and air forces, supported by huge industrial and technological infrastructures. A work that reflected the Napoleonic pattern of operations could hardly retain its former practical value in the new age. Since the First World War Jomini has therefore been known only to students of military history, and their attitude to him, influenced by the legacy of the German military school and by the decline of his influence, has by and large been unfavourable.

Ironically, this attitude largely reflects the legacy of Jomini's own theoretical outlook which was derived from the military thinkers of the Enlightenment. His work has been judged not so much from the

[74] Spenser Wilkinson, *The French Army Before Napoleon*, (Oxford, 1915), 15.

point of view of its success in analysing the warfare of its time but rather for its claim to be a universal theory of war. It has been unfavourably compared with Clausewitz's work, the main themes of which are widely believed to have remained as valid as ever.

Jomini and Napoleon

Jomini's great reputation in the nineteenth century rested partly on the belief that he had revealed the principles of Napoleonic warfare. And Jomini himself was anxious to show that he had not only succeeded in interpreting Napoleon's campaigns but that he was also able to foresee their development in the midst of events, and that his work and talents had been acknowledged by Napoleon, the war-lord himself.

Those who came across the question of the personal and theoretical relationship between Napoleon and Jomini encountered the same historiographical problem; the source of almost all the direct evidence was Jomini himself, who was obviously biased. Furthermore, suspiciously enough, this evidence was produced only in the biography written in 1860 by his friend Lecomte when the persons involved who did not share Jomini's fortune of an extremely long life, were long dead. Much of the biography even appears to be inconsistent with Jomini's own account related throughout his extensive works.

Lecomte's biography alleges, for example, that in 1800 the twenty-one-year-old Jomini, while conversing with friends on the future course of the war, anticipated Napoleon's Italian campaign and manœuvre against the Austrian rear.[75] If this was so, then one must believe that Jomini had predicted Napoleon's plan of campaign and developed his concept of the manœuvre against the enemy's rear even before he began his theoretical studies in 1802–3; and even more improbable, that he, who never failed to emphasize his successes, refrained from mentioning this theoretical achievement in the *Treatise* and *Summary*.

Lecomte's biography also describes Napoleon's impression of the first volumes of the *Treatise* in 1805. After fifty-five years, we are supposed to believe that Napoleon exclaimed (and someone wrote down) the following:

[75] Lecomte, *Jomini*, p. 10.

And people say that the times are not progressing. Here is a young chief of battalion [Jomini's Swiss rank], and of all men a Swiss, who teaches us things that many of my teachers never told me, and that only few generals understand. How did Fouché allow the publication of such a book?! It betrays to the enemy the whole of my system of war![76]

Here too, it is very much out of character for Jomini to have failed to cite much earlier this first-rate praise from Napoleon himself.

On 15 September 1806, on the eve of Napoleon's invasion of Prussia, Jomini prepared a memorandum, cited in his biography, in which he suggested a plan similar to the one actually carried out by the emperor in the Jena campaign.[77] The biography also describes a dialogue which is purported to have taken place between Napoleon and Jomini at the end of a staff meeting. Jomini asked Napoleon if he could join him four days later at Bamberg, implying that he had already deciphered the emperor's secret intentions. The surprised Napoleon asked Jomini who had told him that he was going to Bamberg, and Jomini replied: 'the map of Germany, Your Highness, and your campaigns of Marengo and Ulm'.[78] Although Jomini, closely familiar with Napoleon's previous campaigns, may have anticipated that the emperor would again resort to the manœuvre against the enemy's rear, the story of his conversation with Napoleon raises all the doubts mentioned before. As to his memorandum, Caemmerer has already pointed out that Jomini was well acquainted with the general deployment of the French corps and that by 15 September he probably also had the preliminary marching orders of, at least, his own corps. It was therefore not difficult for him to put two and two together.[79]

Despite what appears almost certainly to be, at the very least, exaggeration on the part of a celebrated military theorist, who in old age, when no one was left to challenge his words, tried to magnify his reputation, there exists a hard core of facts which cannot be ignored regarding his relationship with Napoleon. Jomini, an anonymous foreigner, received from Napoleon in 1805 the rank of colonel and a senior staff position. Even in the socially highly mobile French army this was not common, and the only explanation is that the emperor was genuinely impressed with Jomini's theoretical work. Even if Jomini did not play as decisive a role in Ney's staff as Lecomte's

[76] Lecomte, *Jomini*, p. 29. [77] Ibid. 33–4. [78] Ibid. 47.
[79] Caemmerer, *Strategical Science*, p. 38.

biography would have us believe, it is still very plausible that the talented young man, whose skill for military analysis was undisputed and who had just made a penetrating study of the Napoleonic art of operations, did play a dominant role as chief of staff to Ney (who, as has been repeatedly said, was not particularly known for his intellectual abilities), and was in a position to keep in close touch with Napoleon's operational planning at the time of the events themselves.

Some clues on Napoleon's attitude to Jomini may also be inferred from Napoleon's own remarks and theoretical positions. At St Helena he praised Jomini's account in the *Treatise* of the Italian campaign of 1796–7, and exonerated him from the accusation that he revealed the French war plans to the Allies after his desertion.[80] In one of his talks, reflecting on the benefits that could have been gained from the teaching of Frederick's campaigns in the French military schools, Napoleon said:

> Jomini would have been a good man for that purpose. Such teaching would have put excellent ideas into the heads of young pupils. It is true that Jomini always argues for fixed principles. Genius works by inspiration. What is good in certain circumstances may be bad in others; but one ought to consider principles as an axis which holds certain relations to a curve. It may be good to recognize that on this or that occasion one has swerved from fixed principles of war.[81]

This statement is particularly instructive despite the fact that here Napoleon appears to be in his more sceptical mood. Indeed, his distinction between the role of principles in war and the effect of changing circumstances which are mastered by ingenious inspiration, reveals that Napoleon's theoretical outlook, like Jomini's, was the product of his intellectual background.

That Napoleon, the person who came to symbolize the advent of the nineteenth century, was deeply rooted in the eighteenth century is now widely accepted. Colin had uncovered the influence of contemporary military thinkers on the formation of the military ideas of the young Napoleon, and Napoleon's categories of thought were no exception. After all, Montecuccoli, Feuquières, Folard, de Saxe, Frederick the Great, Guibert, and Lloyd were among the military

[80] C. Montholon (ed.), *Mémoires pour servir à l'histoire de France sous Napoléon, écrits à Sainte Hélène* (Paris, 1823), i. 1.

[81] G. Gourgaud (ed.), *Talks of Napoleon at St. Helena* (London, 1904), 215.

thinkers of the Enlightenment whose works he read in the course of his military education.[82]

Though Napoleon left no theoretical military work and his sporadic dictums were mostly compiled from his dictations to his adjutants during his exile at St Helena, his theoretical outlook is clear enough. The following words could have been written by any of the above-mentioned thinkers as well as by Jomini:

> All great captains have done great things *only* by conforming to the rules and natural principles of the art; that is to say, by the wisdom of their combinations . . . They have succeeded only by thus *conforming*, whatever may have been the audacity of their enterprises and the extent of their success. They have never ceased to make war a *veritable science*. It is only under this title that they are our great models, and it is only in *imitating* them that one can hope to approach them.[83]

The following passage could also have been taken from Jomini's own works:

> Gustavus Adolphus, Turenne and Frederick, as also Alexander, Hannibal and Caesar have all acted on the same principles. To keep your forces united, to be vulnerable at no point, to bear down with rapidity upon important points—these are the principles which insure victory.[84]

This passage indicates that the remarkable similarity in outlook between Napoleon and Jomini encompassed not only their theoretical and historical premises, but also their military conceptions, a fact already pointed out by Caemmerer.[85] 'An army should have but a single line of operations,' wrote Napoleon; 'to operate upon lines remote from each other and without communications between them, is a fault . . . It ought then to be adopted as a principle that the columns of an army should be always kept united, so that the enemy cannot thrust himself between them.' That is because 'by concentrating his forces he may not only prevent their junction but also defeat them one by one'.[86]

Napoleon' maxims, despite their sporadic nature, show clearly that Jomini not only formulated a very penetrating conceptualization of the Napoleonic art of war, but also that he did so in terms very akin

[82] Colin, *L'Éducation militaire de Napoléon*, see esp. ch. I.
[83] My emphases; Napoleon, *Military Maxims* in T. Phillips (ed.), *Roots of Strategy*, maxim no. 112; see also no. 5.
[84] Ibid., no. 77.
[85] Caemmerer, *Strategical Science*, p. 37.
[86] Napoleon, *Military Maxims*, nos. 12, 11, 4.

to those used by Napoleon. Indeed, it is even possible that Napoleon adopted in his dicta some of Jomini's formulations. This affinity is particularly interesting not merely because it supports Jomini's claims. It also indicates that while new intellectual trends devised a new theoretical outlook in which the general's genius, modelled on Napoleon, was given the major role, Napoleon himself—like Jomini and most of his contemporaries—viewed and interpreted war and his own activities as a general through a single conceptual framework, the one propounded by the military thinkers of the Enlightenment.

Part Two

THE GERMAN MOVEMENT CLAUSEWITZ AND THE ORIGINS OF THE GERMAN MILITARY SCHOOL

5

The Reaction against the Enlightenment

New Perspectives on Military Theory

I THE EMERGENCE OF A NEW CLIMATE OF IDEAS

One of the most striking impressions in reading the works of the military thinkers of the Enlightenment is the all-embracing uniformity of their theoretical outlook. They differed, to be sure, in many other respects; for example, their spheres of interest varied and underwent considerable change, and, above all, they were deeply divided on their actual military outlook and ideas. However, they did not differ in the fundamentals of their guiding objective—the search for a general theory of war—which derived from their intellectual environment. Here too there were, of course, varying interpretations and emphases, but the central themes of this objective were both clear and indisputable. War, like all fields of nature and human activity, was susceptible to a comprehensive and systematic theoretical study. In part, it could be reduced to rules and principles of universal validity and possibly even mathematical certainty, for which Newtonian mechanics set the example. However, like the arts, it was also partly in flux, constantly changing, dependent on circumstances, affected by the unforseen and incalculable, and therefore always requiring application through the general's creative genius.

A conspicuous and highly significant fact is that for at least fifty years, from the appearance of Puységur's *Art de la guerre par principes et par règles* in 1748 until the end of the 1790s, virtually no theoretical challenge compromised the domination of this outlook. Nothing is more indicative of its power and close affinity to the highly influential intellectual environment from which it emerged.

This remarkable intellectual coherence came to an end with the appearance of Berenhorst's *Reflections on the Art of War* in 1796–9.

This is not to say that the ideas of the Enlightenment then lost their influence. On the contrary, as we have seen, at the beginning of the nineteenth century, Bülow, Charles, and Jomini developed these ideas in new and highly influential directions; and the overwhelming majority of contemporary military thinkers continued to view war through the perspectives set by the military school of the Enlightenment. However, the absolute hegemony that this school had maintained over military theory was irreversibly broken. Within a few years, Clausewitz began to formulate the most comprehensive and sophsticated expression of new ideas in the field of military thought, thus laying the intellectual foundations for what was to be a new German military school.

The breach in the hegemony of the military school of the Enlightenment, like its fifty years of domination, can only be understood against the background of the general intellectual developments in Europe in the late eighteenth and early nineteenth centuries. It can even be said that only a small minority of the principal themes propounded by Berenhorst and Clausewitz originated within the military field itself. Most were extracted from, and set in motion by, the ideas and ideals of new and powerful cultural trends.

These trends, it must be stressed, were far from forming a single intellectual framework. They expressed a wide variety of views and beliefs which derived from very different and remote sources, represented diverse human groups and inclinations, and aimed at different, if not entirely opposite goals. They were far more heterogeneous than the intellectual framework of the Enlightenment against which they reacted in varying degrees. Indeed, some of these trends were closer to the Enlightenment than to each other.

The diversity is already apparent in the fact that there is no comprehensive name to describe all these trends as a single movement. Irrationalism, historicism, critical philosophy, religious revivalism, vitalism and wholism, idealism, romanticism, conservatism, nationalism, and reactionaryism, were major themes in the new intellectual climate, but none of them could represent the whole. Two accepted terms appear, however, to describe in the most suitable fashion both the general and military points of view. The first is the 'Counter-Enlightenment'. Here too, however, the name must not obscure the fact that the various trends in this cultural movement differed in their antagonism to, and reaction against, the Enlightenment, and, also that most of them were, in fact, heavily in its debt. Furthermore, the emphasis on the negative aspect of the movement may overshadow its clear positive messages. Another accepted term, the 'German Movement', focuses on the country where these trends broke out in the most powerful, diverse, and fertile manner, and had the most profound and wide-ranging influence. It particularly fits the military sphere where the new trends appeared almost exclusively in Germany.

The diversity of the trends in the new intellectual climate was also manifest in the field of military thought, which influenced Berenhorst and Clausewitz in very different ways. On the whole, Berenhorst is a classical exponent of the 'Counter-Enlightenment', whereas Clausewitz may best be understood in the framework of the 'German Movement'. In view of this diversity of ideas and influences, it may be better, perhaps, first to delineate briefly those themes and trends in the new intellectual paradigm whose role in the emergence of the German military school was particularly dominant. This survey is, necessarily, somewhat superficial, and focuses on Germany, on the trends that were critical of the Enlightenment, and the themes that were particularly relevant to the development of military thought. The theoretical outlook of the military exponents of these new trends, Clausewitz in particular, will then be examined separately, in a more extensive, concrete, and distinctive form.

The new cultural trends emerged in Germany in two major waves. The first emerged in the 1770s at the zenith of the Enlightenment in Germany. It was oppositional in nature, associated with a group that operated, to a large extent, outside and against the cultural establishment, and whose most notable members included Hamann, Herder, the young Goethe and the writers and dramatists of the 'Storm and Stress' movement, Lavater and Möser. The second wave emerged throughout Europe at the turn of the century and was accelerated by the threat posed by the French Revolution and the Napoleonic Empire. Sweeping in influence, it embraced the major trends of romanticism, nationalism, and idealism. In addition, in the midst of the German cultural community and inside the fortress of the Enlightenment, Kant exerted an all-encompassing influence, and his decisive role in creating the new intellectual climate was both unique and ambivalent. While he stood, from the point of view of his intellectual development, personality, and self-consciousness, at the pinnacle of the German *Aufklärung*, and was appalled by many of the ideas of the Counter-Enlightenment, he also undermined some of the central doctrines and beliefs of the Enlightenment in his critical philosophy.

The new German Movement challenged the fundamentals of the Enlightenment world-view, which may be considered under four major headings: the conceptions of knowledge and reality, man, art, and history. Its opposition to the French intellectual and political imperialism went hand in hand with the awakening of the German national sentiments which developed in a highly political direction, placing a strong emphasis on the role of the state.

(*a*) Behind the intellectual world of the Enlightenment there stood the tradition of natural law, the legacies of both Cartesian rationalism and British empiricism, the neo-classical conceptions in the arts, and the model of Newtonian science. The conception of knowledge consolidated from these sources by the men of the Enlightenment implied that, essentially, the complex world of experience was governed by relatively few principles which were at once simple, fundamental, universal, and tending to precision. Newton's three laws of mechanics exemplified all these qualities most remarkably.

By contrast, the thinkers of the Counter-Enlightenment regarded this conception of knowledge and reality to be fundamentally false or, at least, highly exaggerated. The world was for them not basically simple but, on the contrary, highly complex, composed of innumerable and unique elements and events, and always in a state of flux. Hence their much cooler attitude to the scientific ideal embodied in Newtonian science.

Hamann (1730–87), Kant's rebellious disciple and the spiritual mentor of the men of the 'Storm and Stress' period, scorned the Enlightenment's blindness to, and loss of touch with, rich and vital reality on which it arrogantly attempted to force artificial, crude, and superficial principles and conceptual frameworks. Genuine knowledge was always the knowledge of singular and unique cases. The sciences, which Hamann treated with contempt and in which he was hardly interested, could, at best, serve as crude approximations in resolving certain practical, fundamentally mechanical problems.[1]

[1] For Hamann's life, world-view, and influence see W. M. Alexander, *Johann Georg Hamann, Philosophy and Faith* (The Hague, 1966). For a penetrating, concise outline see Isaiah Berlin's marvellous articles, esp. 'Hume and the Sources of German Anti-Rationalism' in his *Against the Current: Essays in the History of Ideas* (Oxford, 1981), pp. 165–70; and also 'The Counter-Enlightenment', ibid. 6–9, to which this chapter owes a great deal.

Hamann's friends among the men of the 'Storm and Stress' period shared his criticism of the Enlightenment for its totally erroneous attempt to force the categories which had proved successful in physics on reality as a whole. Despite his admiration for the achievements of the natural sciences, Herder (1744–1803) believed that their conceptions, while suitable for the inanimate and simple bodies of mechanics, were totally unsuitable for understanding other spheres of a rich and complex world, in particular for the understanding of man and society.[2]

Goethe, who was enthusiastically interested and actively involved in the scientific developments of his time, believed that the analytic and classifying method did not even suit the natural sciences. Already in the Enlightenment, the biological and vitalistic theories of Maupertuis and Buffon had aroused much interest in Diderot, who looked upon the domination of mechanics over his period with some apprehension.[3] The discoveries in the fields of electricity and the chemistry of gases, during the last third of the eighteenth century, further reinforced the tendency to view nature through organic and vitalistic conceptions. Goethe pointed out that the classifications of biology and mineralogy were imposing human conceptual frameworks on a nature whose diversity of forms and changes was infinite. The long list of 'intermediate cases' and 'exceptions' created by these classifications revealed their artificiality all too clearly. In his diatribe against Newton's optics, Goethe argued that any attempt to base the diversity of the spectrum of colours on the crushing of the white colour was to be totally rejected.[4]

Paradoxically, Kant's all-embracing influence also worked to restrict the belief in the power of reason and to weaken the domination of the model of Newtonian science. Though one of the declared aims of the *Critique of Pure Reason* (1781) was to rescue the achievements of the natural sciences from the threat of scepticism, it only achieved this by excluding whole sections of reality from the

[2] For Herder's attitude to science, see H. B. Nisbet, *Herder and the Philosophy and History of Science* (Cambridge, 1970).

[3] On the latent tension within the Enlightenment between the domination of the mechanistic explanation and the organic-vitalistic view, see Colin Kiernan, 'Science and the Enlightenment in Eighteenth Century France', in T. Besterman (ed.), *Studies on Voltaire and the Eighteenth Century*, LIX (1968).

[4] On Goethe and contemporary science, see George A. Wells, *Goethe and the Development of Science* (The Netherlands, 1978).

domain and capacity of reason. Moreover, the aim of the *Critique of Practical Reason* (1788) was to establish the autonomy of the human soul from the regularity which dominated nature.

These developments also found expression in the works of the early Romantics at the turn of the century, and were philosophically formulated in Schelling's *Naturphilosophie*. Nature embraced an endless diversity of forms, was motivated by vitalistic forces, and maintained a dynamic relationship with man.[5] Attempting to remedy what they regarded as ruptures created by Kant's philosophy between the various faculties of man and between man and reality, the idealists Fichte, Schelling, and particularly Hegel, developed to the utmost the holistic and integrative notions inherent in the German Movement. All elements of reality were but aspects and manifestations of a single whole.

These new perspectives had particular bearing on the study of man, the real interest of the men of the German Movement.

(*b*) The attitude of the men of the German Movement to the legacy of the British empiricists in human psychology was ambivalent. On the one hand, they admired the achievements of empiricism in describing the construction of human consciousness from the materials of experience, and the primacy they gave to the study of man in understanding the world. On the other hand, following Leibnitz and Kant, they rejected the empiricist claim that man was a *tabula rasa*, and the dissection of the human mind into atomistic impressions and sensations. This conception, they felt, missed the essence of man as an active, creative, and imaginative unity which integrated the impressions of experience. The deep and multifaceted human experience, as intuitively and intimately known to every individual, was diametrically opposed to the crude, mechanistic, and skeletal system portrayed both by associative psychology and the materialists. Goethe expressed the attitude he shared with his friends when he called Holbach's work 'ghostly' and 'corpse-like'.[6]

The men of the Counter-Enlightenment were interested in direct and concrete human experience. This orientation was deeply rooted in, among others, the pietist stream of Lutheranism whose influence

[5] See esp. Alexander Gode von Aesch, *Natural Science in German Romanticism* (New York, 1941).

[6] Goethe, *Dichtung und Wahrheit*, Bk. II, in *Werke*, x. 537–9 (Zurich, 1949–52); cited by Roy Pascal, *The German Sturm und Drang* (London, 1953), 131.

from the end of the seventeenth century, particularly in East Prussia, was considerable. The works of Hamann, Lavater, Herder, Jacobi, and even Kant were deeply embedded in this powerful spiritualist tradition. The pietist emphasis on personal experience and its suspicion of all dogma had at first, paradoxically, helped to pave the way for the Enlightenment in Germany, but were later directed against the Enlightenment, against the tyranny of its ideas, and both its atheism and rationalist natural religion.[7]

Hamann, the major exponent of spiritualism, argued that only imaginative, empathic insights, rather than abstract and universal principles, could penetrate into the wealth and uniqueness of human reality. Man was a complete creature, whose whole personality, rather than narrow aspects of it, were expressed in all spheres of his activity. Goethe summarized Hamann's teaching in saying: 'Everything that man undertakes whether it be produced in action or word or anything else, must spring from his whole united powers; all separation of powers is to be repudiated.'[8] He and other 'Storm and Stress' writers, like Merck, Lenz, and Klinger, highlighted man's vitality, activity, and power of feelings in their plays. The men of the movement enthusiastically accepted Rousseau's human sensitivity and, as we shall see, the emphasis that the British aesthetic school, which originated with Shaftesbury, had put on the creative imagination.[9]

At the turn of the century all these themes were raised to prominence in the works of the early Romantics—the brothers Schlegel, Tieck, and Novalis. Their friend, the celebrated and influential preacher and religious thinker Schleiermacher, also stressed the uniqueness and potential of feelings, sensations, and thoughts revealed in every individual. He gave these ideas systematic expression in his *Monologen* (1800). The Romantics' philosophical patron Fichte, in his *Science of Knowledge* (1794) and *Theory of Knowledge* (1797), made man the creator of reality through his free spiritual activity. And the Romantics adapted this to promote the

[7] For this well-known relationship, see e.g. K. S. Pinson, *Pietism as a Factor in the Rise of German Nationalism* (New York, 1968).

[8] Goethe, *Dichtung und Wahrheit*, XII, in *Werke*, x. 563; cited by Pascal, *Sturm und Drang*, pp. 9–10.

[9] For the intellectual world of the men of the 'Storm and Stress' period, see Pascal's learned *Sturm und Drang*.

omnipotence of the creative imagination and force of feelings in the arts.[10]

(*c*) The emphasis on the creative, unique, and imaginative character of the individual, which could not be reduced to abstract and mechanical principles, was closely associated with a growing reaction against the legacy of seventeenth-century neo-classicism in the arts. In Britain, the country least influenced by neo-classicism, a line of writers and critics had been following in Shaftesbury's footsteps from the beginning of the eighteenth century. Leibnitz introduced Shaftesbury's influence to the continent, and particularly to Germany. There, the tenets of neo-classicism were directly challenged in the Gottsched affair, while Diderot represented moderate criticism of neo-classicism in France. All these people promoted the ideas of the originality of genius and the force of creative imagination and used them to counter the conceptual frameworks of neo-classicism. In his *Critique of Judgement* (1790) Kant consolidated the transformation in the eighteenth-century outlook on artistic creation, and the Romantics gave the last great push to the decline of neo-classicism.[11]

The reaction against neo-classicism went hand in hand with a powerful wave of interest in and admiration for forms of art hitherto considered by the men of the Enlightenment to be barbarous, lacking in taste or aesthetic knowledge, and produced by uncivilized or semi-civilized societies. The enthusiasm for Homer; for the poetry, folk-tales, and myths of the ancient Hebrews, Celts, and Germans, and of primitive people in general; for the spirit and art of the Middle Ages; and for the expressive genius of Shakespeare, whom the neo-classicists considered with horror to be a barbarous and demonic writer who disregarded all genres and

[10] For a general and critical survey of German Romanticism, see Ralph Tymms, *German Romantic Literature* (London, 1955); and for the Romantic's world-view see H. G. Schenk, *The Mind of the European Romantics* (London, 1966).

[11] In addition to the works cited in Ch. 2 n. 7 see esp. James Engell's highly comprehensive *The Creative Imagination, Enlightenment to Romanticism* (Cambridge Mass., 1981); see also L. W. Beck, *Early German Philosophy, Kant and His Predecessors* (Cambridge Mass., 1969), 278–88; and, for a defence of neo-classicism against stereotyped criticism, see E. B. O. Borgerhoff, *The Freedom of French Classicism* (Princeton, 1950). 17th-cent. neo-classicism as a conception of art is obviously not to be confused with late 18th-cent. German *Klassizismus* as an artistic style and view of life, mainly associated with the Weimar poets. The concept has been used in different ways for different periods.

conventions, were all closely linked to a profound transformation in viewing the past.

(*d*) In the second half of the eighteenth century a new historical outlook, later to be known as historicism, began to take shape. Criticism was levelled against the tendency of the men of the Enlightenment to view other societies and historical periods through the perspectives and values of their own time, which were thus perceived as a universal standard of measurement for the interpretation, criticism, and rejection of complete historical eras.

The beginnings of this transformation are to be traced, however, to the Enlightenment itself, particularly to the influence of Montesquieu, who introduced his contemporaries to a new depth of analysis of the relationships between the environmental, economic, religious, political, and constitutional factors which moulded the diversity of societies and cultures. Rousseau's yearning for primitive man, reflecting a growing alienation from modern society, was also highly important. These influences were reinforced, as mentioned above, by a wave of interest, particularly in Britain and Germany, in the ancient Greeks, biblical Hebrews, archaic peoples of the North, and Gothic architecture. All this influenced, and culminated in, the work of Herder.[12]

Herder argued in his works, particularly *Auch eine Philosophie der Geschichte* (1774) and *Ideen zur Philosophie der Geschichte der Menschheit* (1784–91), that every culture was a unique historical entity that stemmed from the particular circumstances and experience of its time and place and, in turn, expressed them in the totality of its values, ways of life and thought, institutions, and creative art. A dogmatic examination according to so-called universal standards precluded any real understanding, which could only be achieved by sympathetic and imaginative insights into the concrete conditions of a bygone reality and consciousness, aiming to reconstruct them in their own particular terms. Rather than superficial abstractions, a close and detailed study of the diverse forms of specific historical

[12] For the diversity of sources which influenced the genesis of historicism, see particularly the study of one of the most distinguished exponents of the movement: Friedrich Meinecke, *Historicism, the Rise of a New Historical Outlook* (London, 1972); for a more critical approach see G. Iggers, *The German Conception of History* (Middletown Con., 1968). A recent comprehensive study is P. H. Reil's *The German Enlightenment and the Rise of Historicism* (Berkeley, 1975).

situations was needed.[13] As an example of such a study, the men of the 'Storm and Stress' period were delighted with Justus Möser's close, penetrating, and vivid records of the ways of life, customs, and affairs of his fellow townsmen and peasants in the small principality of Osnabrück, and with his research into their medieval past.[14]

In Strasburg, Herder revealed to the young Goethe the beauty of the city's medieval cathedral, built in the Gothic style that was despised by the men of the Enlightenment. The writers of the 'Storm and Stress' movement felt an affinity to the past, and the Romantics enthusiastically embraced its diversity. Folk-songs and folk-tales, regarded as a vulgar subculture by the men of the Enlightenment, were elevated by Herder to the status of creative, authentic, and revealing indications of past ways of life. The Romantics followed in his footsteps not only in their literary themes but also in compiling folk-songs and legends. Clemens Brentano and Achim von Arnim published in Heidelberg an anthology of German folk-songs, *Des Knaben Wunderhorn, Alte Deutsche Lieder* (1805–8). The Grimm brothers followed suit with their celebrated collection of folk-tales, *Kinder und Hausmärchen* (1812).

These were the beginnings of historicism whose influence on the sciences of man was revolutionary and all-embracing. Human reality, according to the historicist message, was moulded by history, and changed with time and place, thus undermining any universal generalization. It could genuinely be understood only in a particular historical context. Directed against French Revolutionary ideas, this message, bolstered by Burke's highly influential *Reflections on the Revolution in France* (1790), was widely voiced by political theorists. Adam Müller in *Die Elemente der Staatskunst* (1809) lay the foundation of the historical school of economics. The principles that the political economists of the eighteenth century (headed by Adam Smith) had formulated and regarded as the universal rules of economics were considered by this school to be a reflection of the

[13] For Herder's conception of history see, in addition to the works cited in the previous note, A. O. Lovejoy, 'Herder and the Enlightenment Philosophy of History', in his *Essays in the History of Ideas* (Baltimore, 1948), 166–82; G. A. Wells, 'Herder's Two Philosophies of History', in the *Journal of the History of Ideas*, XXI (1960), 527–37; and I. Berlin, 'Herder' in id., *Vico and Herder* (London, 1976).

[14] See Knudsen, *Justus Möser and the German Enlightenment*; for a good concise account see Klaus Epstein, *The Genesis of German Conservatism* (Princeton, 1966), ch. 6.

particular conditions and interests prevailing in the capitalist, proto-industrial Britain of the time. Friedrich Karl Savigny launched the historical school of jurisprudence in his celebrated *Vom Beruf unserer Zeit für Gesetzgebung und Rechtswissenschaft* (1814). Law was not, and could not be, determined according to universal and abstract principles, argued Savigny; it developed out of the particular historical conditions of every society. Schleiermacher presented the dogmas, conventions, and institutions of religion as changing throughout history. And Hegel bonded the human mind and philosophy to history which reflected the various stages in the development of consciousness. Finally, in the more strictly historical field, there emerged the great historical school of the nineteenth century, associated with the name of Ranke.

These were some of the major themes in the German reaction to the dominating ideas of the French Enlightenment.[15] From being a provincial and somewhat backward culture which Möser—to the delight of his friends—defended in his *Über die deutsche Sprache und Literatur* (1781) against the scorn of the French-oriented Frederick the Great, German culture in the last decades of the eighteenth century became the centre of stimulating intellectual activity. Its growth was therefore linked with an anti-French tendency and awakening national sentiments. A German cultural self-awareness emerged in reaction against French intellectual imperialism, and developed, in response to Napoleonic political imperialism, in a clear political direction with a strong emphasis on the primacy of the state.

[15] For the social and economic aspects of the transition, see Henri Brunschwig, *Enlightenment and Romanticism in Eighteenth Century Prussia* (Chicago, 1974). A very critical assessment of the origins of German cultural identity in the 18th cent. is to be found in chs. 4 and 6 of Arnold Hauser's breath-taking *The Social History of Art*, ii (London, 1951).

II BERENHORST: COUNTER-ENLIGHTENMENT AND THE CRITICISM OF THE FREDERICKIAN SYSTEM

As pointed out by Berlin, the fierce and wide-ranging opposition to the ideas of the Enlightenment was as old as the Enlightenment itself. This opposition came, however, from traditionalist and conservative elements outside the intellectual circles with whom they had no common ground for genuine communication. This situation changed with the Counter-Enlightenment. The opposition now came from within the intellectual élite, had developed from the legacy of the Enlightenment itself, and challenged its ideas in its own language.[1]

This picture holds equally true in the military field. As mentioned earlier, many soldiers were probably alien to the 'military Enlightenment', and perhaps still more were simply indifferent to it. But naturally, very few, such as Brenckenhoff in his *Paradoxa*, gave their thoughts or feelings systematic intellectual and literary expression, and thus the absolute domination of the ideas of the Enlightenment over the field of military theory was hardly compromised. This changed with the work of Georg Heinrich von Berenhorst (1733–1814). Typical of the men of the Counter-Enlightenment, such as Hamann, he was a child of the Enlightenment who underwent a profound intellectual and psychological transformation. Experiencing the religious-spiritualist revival and influenced by Kant's critical philosophy, he adapted the new intellectual trends to the military field in a sophisticated, sometimes sardonic, sometimes aphoristic manner.

Berenhorst was the illegitimate son of Prince Leopold I of Anhalt-Dessau, the famous *Alte Dessaur*, one of the architects of the Prussian army and young Frederick's right-hand man. At the age of fifteen he joined an infantry regiment, and as a member of Prince Heinrich's staff and, from 1759, in Frederick's own headquarters, he took part in the great campaigns of the Seven Years War. He then entered the diplomatic service of his native

[1] Berlin, 'The Counter-Enlightenment', in *Against the Current*, p. 1.

principality, and after retirement in 1790, embarked upon his literary career.[2]

It is, however, Berenhorst's intellectual and psychological development, more than the biographical outline sketched above, that is of particular interest. Berenhorst himself left us an autobiographical essay which has not received the attention it deserves. He entitled it 'Selbstbekenntnisse', 'Confessions', in direct reference to the celebrated work of Rousseau, the hero of the new appeal to emotions and the inner world.[3]

In his youth, wrote Berenhorst, he was close to religion, but as he grew up his attitude changed. He read Helvétius's *De l'esprit*, lost his faith, and became a materialist. He accepted the explanation of man as a machine, and his religion was 'pantheism without morality'. He went on to read Lucretius, the exponent of atomism and materialism in antiquity, and the writings of Montaigne, Bayle, and Voltaire, who promoted scepticism, the critical spirit, and religious toleration. He was deeply influenced by the works of Nicolas Fréret, the leading figure in the French Academy of Inscriptions in the first half of the eighteenth century, who laid the foundations for the historical criticism of Christianity.[4]

Then, in his late thirties and forties, came the great spiritual transformation. He read the works of authors such as N. S. Bergier, one of the devout participants in the *Encyclopédie*, J. F. W. Jerusalem, one of the neologians, A. von Haller, the celebrated scientist and poet, and Lessing, who all strove to eliminate the conflicts between revealed religion and reason.[5] However, he was primarily influenced by the major exponents of the great pietist, spiritualist revival. He watched with interest Lavater's onslaught on Mendelssohn (1769), and the famous controversy between the latter and Jacobi regarding the nature of Lessing's religious faith (1785). He came to reject both natural law and natural religion, and to

[2] A concise biography of Berenhorst and an account of his works is contained in Edward Bülow's introd. to a collection from Berenhorst's literary remains: *Aus dem Nachlasse* (2 vols.; Dessau, 1845 and 1847). Many other items from Berenhorst's family archives are cited in Rudolf Bahn's *Georg Heinrich von Berenhorst* (doct. diss.; Halle, 1911). Eberhard Kessel's 'Georg Heinrich von Berenhorst', in *Sachsen und Anhalt*, IX (1933), 161–98, is a perceptive analysis of Berenhorst's work. Also see E. Hagemann, *Die deutsche Lehre vom Kriege; von Berenhorst zu Clausewitz* (Berlin, 1940), 6–20.

[3] See Berenhorst, 'Selbstbekenntnisse', in E. von Bülow (ed.), *Aus dem Nachlasse*, ii. 3 for the title's reference to Rousseau.

[4] Ibid. 4–5.

[5] Ibid. 6–7.

promote inner life, intuition, emotions, and free will.[6] It is no coincidence that his literary remains include several critiques of plays of the 'Storm and Stress' dramatists, particularly the 'great Goethe'.[7]

Kant's influence was equally decisive. In her perceptive portrayal of German culture, *De l'Allemagne* (1813), IV, 112, Mme de Staël described the all-embracing effect of his work: 'the *Critique* [of Pure Reason] created such a sensation in Germany that almost everything achieved since then, in literature as well as philosophy, derives from the impetus given by this work'. The military field was no exception. In his 'Confessions' Berenhorst wrote that he laboured much to understand Kant's works and succeeded in gaining access to his speculative philosophy. Kant saved free will and set the boundaries of human knowledge. Berenhorst regarded his own work to be, to some extent, a Kantian critique of military theory.[8]

The first volume of Berenhorst's *Reflections on the Art of War, its Progress, Contradictions, and Certainty* appeared in 1796 while the second, together with a revised edition of the first, appeared in 1798, and the third in 1799. According to a contemporary, at that time 'no book was as widely read as *Reflections*'.[9]

Berenhorst's historical account of the development of the science and art of war (that is, the intellectual treatment of war as embodied in military institutions and military organization) clearly reveals his heavy debt to the Enlightenment. As a face-to-face encounter, classical warfare, he wrote, was based on courage and physical strength. Yet the ancients also brought the science of war to a pinnacle of perfection which had not been achieved since. 'The ancients', he stated, 'I mean the Greeks and Romans, were, compared with the moderns, how should one put this?—more artistic.'[10] The Middle Ages were in this respect very different, and there was very

[6] 'Selbstbekenntnisse', *Aus dem Nachlasse*, ii. 6–14. Berenhorst's attitude to natural religion is also revealed in several references in his major military works; see e.g. *Betrachtungen über die Kriegskunst, über ihre Fortschritte, ihre Widersprüche und ihre Zuverlässigkeit* (3rd edn., Leipzig, 1827), 170.

[7] *Aus dem Nachlasse*, ii. 131 ff.

[8] 'Selbstbekenntnisse', *Aus dem Nachlasse*, ii. 14–16; again, in a typical aside, see e.g. Berenhorst's discussion of the distinction between *Vernuft* and *Verstand*: *Aphorismen* in *Betrachtungen*, pp 539–40.

[9] Cited in Jähns, *Geschichte der Kriegswissenschaften*, p. 2128.

[10] Berenhorst, *Betrachtungen*, pp. 40–1; for war in the classical period and a companion between the ancients and the moderns, see chs. I and V.

little to be said about them. As they constituted an era of backwardness resembling the dawn of warfare before the classical period, 'courage and physical force alone' decided the fate of wars.[11] The art and science of war resumed development with the military reforms and innovations of the Dutch during their wars of independence, and of Gustavus Adolphus.[12] Louvois's military organization under Louis XIV gave the French hegemony in the science of war for a hundred years,[13] while a new level of achievement was reached by Frederick's Prussia.

Yet military science and art, which were the intellectual parts of war, were different in nature from what they had been assumed to be by the military thinkers of the Enlightenment. Indeed, other factors may have been of far greater importance in war. In a note entitled 'The Main Idea of the Whole Work' written when he was composing the *Reflections*, Berenhorst stated that the art of war, like the rest of the sciences and the arts, advanced knowledge and supported innate talent. However, it was not based on immutable laws but was rather associated with the unknown and uncontrollable modifications of the human spirit, and operated in an environment saturated with will-power and emotions.[14] In connection with the attacks launched by the Prussian army against all odds in the battles of the Seven Years War, contradicting 'all the rules of the art', Berenhorst wrote that 'if at that moment someone, such as, perhaps, Puységur, had flown above the belligerents in a balloon, he would have said: "I judge according to the principles—the Prussians must be beaten and defeated". But fate was different.' The spirit of the army and blind chance carried the day. 'The Prussians won in spite of the art.'[15]

According to Berenhorst, the moral forces that animate the troops are therefore a major factor in the conduct of war. Far from being automata, the troops could be inspired with a fierce fighting spirit, particularly when motivated by patriotic enthusiasm. Indeed, Berenhorst was the most respected critic of the Frederickian system in the great military debate in Germany over French Revolutionary warfare. His criticism derived, however, from much deeper roots, reflecting an older and more comprehensive opposition. It was typical of the men of the Counter-Enlightenment who detested the 'King

[11] Ibid., p. 9. [12] Ibid., ch. III. [13] Ibid., ch. IV.
[14] *Aus dem Nachlasse*, i. 3. [15] *Betrachtungen*, pp. 66–7.

of Prussia' with his bureaucratic, lifeless, 'machine-like' state, and French rationalistic orientation. Frederick was severely criticized by Berenhorst for lacking national consciousness, and assimilation into a foreign culture.[16] It was not surprising that the king, who could barely speak German, and whose people were but subjects to him, regarded his troops as no more than soulless material for his war-machine, and had no appreciation for the military potential of patriotic sentiments. All the interrelated elements of the Prussian military system—its mercenary troops, ruthless discipline, mechanical drill, and linear tactics—suppressed rather than enhanced moral forces. Armies should reintroduce the pike, as de Saxe recommended, and rely on shock tactics to achieve decision in battle.

A critique of Berenhorst's theoretical views and a defence of the Frederickian system, *Betrachtungen über einige Unrichtigkeiten in den Betrachtungen über die Kriegskunst* (1802) was written by the military scholar and *Aufklärer* Colonel Massenbach, who was a contributor to Nicolai's *Allgemeine deutsche Bibliothek*, the literary bastion of the Berlin Enlightenment, and whose career was later ruined by the defeat of 1806. Berenhorst replied in the same year with a polemic work which stressed the message of the *Reflections* even more, and he reasserted his ideas in *Aphorisms* (1805).

War, he wrote, unlike mathematics and astronomy, could not be formulated as an a priori science.[17] He emphasized his affinity to and belief in the sciences, but requested his critics to bear in mind the numerous examples in military history in which armies with natural courage, ignorant of the art of war, had carried the day, and the many others in which principles had been revealed as useless or inadequate. 'What then is left of the certainty, let alone usefulness, of science?'[18] Rules and principles tend to be artificial, dogmatic, and uncircumstantial; principles, abstracted from experience, are indiscriminately applied to an altered situation. 'What is the use of rules when one is covered up to one's ears with exceptions?'[19] The emphasis on the science and art of war corresponds to the old illusion of the philosophers about the intellectual essence of man.[20] In fact, the real power of armies rests in the moral and physical force of the troops rather than in all the sciences of the officers.[21] The qualities

[16] See e.g. *Betrachtungen*, p. 170. [17] *Randglossen* in *Betrachtungen*, p. 477.
[18] Ibid. 472–3. [19] Ibid. 499–500. [20] Ibid. 477. [21] Ibid. 449–50.

and characteristics of a general are mainly innate which the sciences can develop only slightly, though they provide him with ideas—particularly the study of military history and the art of war—and they improve him as a human being.[22]

Berenhorst's writings in 1802–5 indicate a growing shift from a critical approach to pronounced theoretical scepticism. The developments in both his military and intellectual environment undoubtedly contributed to this. In the wake of Prussia's defeat by Napoleon in 1806, Berenhorst played bitterly with several variations on the ironic pun: 'the French and Prussian generals divided the art of war between them; the Prussians took the former and the French the latter'.[23]

In relation to Jomini's principles which he regarded as fundamentally sound, Berenhorst in 1809 employed the argument he had already used in *Reflections* concerning the art of war of antiquity. Though the Greeks and Romans had subjected war to the highest level of intellectual control, he wrote then, their science of war had played to their advantage only as long as they confronted barbarous peoples; when they fought each other science had been neutralized, and courage and talent had again decided the issue. Now, the same applied to Jomini's principles. As long as Napoleon was the only one to exercise them, he could achieve success, but once everyone employed his system, it would cancel itself out, and numerical superiority, courage, and the general's fortunes would again reign supreme.[24] Theoretical argument aside, this was a penetrating anticipation of the events of 1813–15.

Responding to a letter in which Valentini had told him that Clausewitz did not believe in a general art of planning operations, Berenhorst wrote in 1812: 'I tend to agree with him . . . the [plans] are rendered absurd in one way or another by unforeseen circumstances . . . Then should we proceed without any plan just into the blue? I wish I could reply "yes", but fear of the gentlemen who think in formulae holds me back.'[25] Paradoxically, Berenhorst's affinity to the Enlightenment is strikingly revealed here. He could only see an alternative between a science of principles and anarchy, and despite his theoretical scepticism he could not embrace the latter.

[22] *Aphorismen* in *Betrachtungen*, p. 542. [23] *Aus dem Nachlasse*, i. 192–3.

[24] For the argument in relation to antiquity see *Betachtungen*, p. 2; and for its application to Napoleonic warfare and Jomini's principles see *Aus dem Nachlasse*, ii. 295–6.

[25] *Aus dem Nachlasse*, ii. 333, 353–4; cited by Paret, *Clausewitz*, p. 206.

6

Clausewitz

Demolishing and Rebuilding the Theoretical Ideal

I SCHARNHORST'S PLACE AND LEGACY

Scharnhorst discovered Clausewitz, acted as a second father to him, guided his development, and paved the way for him to reach the upper levels of the Prussian army and state and to be at the centre of the military and political events of the period. Furthermore, Scharnhorst made what was perhaps the most decisive contribution to the formation of Clausewitz's military outlook and theoretical conceptions. This is the view shared by all students of Clausewitz. Clausewitz called him 'the father and friend of my spirit'.[1]

What then was Scharnhorst's outlook on military theory, and what exactly did he bequeath to Clausewitz? These questions have received only cursory treatment. In his political and military views as well as in his work in reforming the Prussian army, Scharnhorst is said to have rejected radicalism from both the right and the left, and to have striven to harmonize the achievements of the *ancien régime* with the innovations and requirements raised by the Revolution.[2] However, this characteristic of his life's work and world-view has not been fully recognized in his approach to military theory. He has been largely portrayed as one who rejected and opposed the theoretical conception spread by the military thinkers of the Enlightenment.[3]

[1] A letter to his fiancée, 28 Jan. 1807; K. Linnebach (ed.), *Karl und Marie von Clausewitz, Ein Lebensbild in Briefen und Tagebuchblättern* (Berlin, 1916), 85.

[2] A picture established by Max Lehmann, *Scharnhorst* (2 vols., Leipzig, 1886–7).

[3] Apart from the studies about Clausewitz, see esp. Höhn's valuable *Revolution, Heer, Kriegsbild*, esp. pp. 467–514, and his more concise *Scharnhorst's Vermächtnis* (Bonn, 1952); see also Hansjürgen Usczeck, *Scharnhorst, Theoretiker, Reformer, Patriot* (East Berlin, 1979), which largely follows in Höhn's footsteps with a Marxist twist and much contemporary rhetoric.

There are two main reasons for this image. Firstly, the general unawareness of the distinctive ideas and form of the military school of the Enlightenment explains why Scharnhorst's extensive literary activity and theoretical conceptions from the 1780s, though not unknown, have mostly been studied from a political and military point of view, while their intellectual context has largely remained obscure. Scharnhorst's link with the intellectual world and with the prominent authors of the eighteenth century as outlined by Stadelmann has thus not been fully appreciated either.[4] Secondly, views about Scharnhorst's theoretical approach have naturally been influenced by what is known about Clausewitz's theoretical outlook. This tendency was reinforced by Clausewitz himself, whose close relationship with Scharnhorst occurred at the beginning of the nineteenth century when the latter was emphasizing a particular aspect of his ideas. Clausewitz too strengthened the impression that, fundamentally, Scharnhorst rejected the traditional conceptions of military theory.

In truth, Scharnhorst was from his youth one of the best-known active military *Aufklärer*s. Throughout his life he on the one hand defended the theoretical vision of the Enlightenment against its opponents, while on the other he rejected the radical interpretations of this vision, particularly when they took a new revolutionary direction at the turn of the eighteenth century.

Gerhard Johann David Scharnhorst was born in 1755 to a retired non-commissioned officer of the Hanoverian army and to a daughter and heiress of an affluent free farmer.[5] In 1773, he entered the military academy founded by Count Wilhelm zu Schaumburg-Lippe-Bückeburg in his tiny state near Hanover, an event that was to mould his entire career and intellectual development.

Count Wilhelm (1724–76), an international soldier and exponent of the Enlightenment, was brought up in England and France and showed a lively intellectual interest in many fields, especially in mathematics and history. He gained his military experience in the

[4] Rudolf Stadelmann, *Scharnhorst, Schicksal und geistige Welt, ein Fragment* (Wiesbaden, 1952).

[5] For Scharnhorst's life story see the monumental biographies of Georg Heinrich Klippel, *Das Leben des Generals von Scharnhorst* (3 vols.; Leipzig, 1869–71), and Max Lehmann, *Scharnhorst*. For a concise account in English, see ch. 4 of P. Paret's *Clausewitz and the State*.

War of the Austrian Succession in Holland and Italy, and in the Seven Years War he commanded the defence of Portugal, Britain's ally, against a Spanish invasion. Among the acquaintances with whom he corresponded and conversed were Mendelssohn, Goethe, Möser, and Herder. Influenced by the writings of Thomas Abbt, who called for the revival of Roman patriotism, he experimented with a citizen militia in his tiny state. The military reforms that he introduced, the book that he wrote, *Mémoires pour servir à l'art militaire défensif* (1775), and the military academy that he established, all reflected the military ideas and ideals of the Enlightenment.[6] The academy's broad curriculum, drawn up by the count himself, who was also the chief instructor, was typical of the military academies and educational programmes of the period. The cadets were taught pure and applied mathematics, civil architecture, physics, natural history, economics, geography, history and military history, and the military sciences of tactics, artillery, and fortifications.[7]

After Count Wilhelm's death, Scharnhorst in 1778 transferred to the Hanoverian service. His interest in military education was now further developed as he collaborated with other officers of similar persuasions in a series of pioneering projects. The commander of the cavalry regiment in which he entered, von Estorff, (himself a notable military *Aufklärer* and author of a book, *Fragmente militairischer Betrachtungen über die Einrichtung des Kriegswesens in mittlern Staaten* (1780)), founded a regimental school for the officers and NCOs where mathematics and military studies were taught. Scharnhorst was an instructor in this school and used the experience for further expanding his military studies and developing his educational ideas. The writings of Nicolai, whom Scharnhorst considered to be the foremost military scholar in Germany, were a major source of influence.[8] In 1782 Scharnhorst was appointed instructor in the newly formed artillery academy in Hanover whose syllabus was again comprised of geometry, pure and applied mathematics, fortifications, artillery, and tactics.[9] In those years he also began his extensive literary activities which soon rendered him one of the best-known figures in the community of the military *Aufklärers*.

[6] Klippel, *Leben*, i. 38–60; Lehmann, *Scharnhorst*, i. 12–29. For his period in Portugal, see C. Harraschik-Ehl, *Scharnhorsts Lehrer: Graf Wilhelm von Schaumburg-Lippe in Portugal* (Osnabrück, 1974).

[7] Klippel, *Leben*, i. 51.

[8] Ibid. 70–3.

[9] Ibid. 84–90.

As mentioned above, from 1782, when he was twenty-seven years old, Scharnhorst initiated and edited a series of military periodicals which soon became among the most widely read of their kind in Germany with hundreds of subscribers. The *Militair Bibliothek* (four issues; 1782–4) and *Bibliothek für Offiziere* (four issues; 1785) mostly contained translated selections from the latest military literature in Europe, but also included an increasing number of articles and critiques. The *Neues Militärisches Journal* appeared in thirteen volumes from 1788 to 1805, with a lull between 1793 and 1797, when Scharnhorst took part in the wars of the Revolution.[10]

Scharnhorst was also the author of two widely circulated military works. The *Handbook for Officers on the Applied Parts of the Sciences of War* was a mine of information on the various branches of war, and included extensive technical and statistical data on the organization and equipment of contemporary European armies—impressive evidence of the scope of Scharnhorst's military knowledge.[11] The more concise *Military Pocket-book for Use in the Field* was a general manual on the conduct of war, with instructions for marches, camps, and reconnaissance; for warfare in the open field, against field fortifications, and during a siege; and for the use of cavalry, infantry, artillery, and engineering: it was a typical product of the military literature of the *Auflkärung*.[12] The book gained much popularity, went through several further editions (1793, 1794, and 1815), and was translated into English (1811). A study of Scharnhorst's extensive writings in these periodicals and books and in other unpublished works elucidates the nature and context of his theoretical outlook.[13]

The young Scharnhorst opened his introduction to the *Militair Bibliothek* (1782) with the proclamation, typical of the military

[10] The six vols. which appeared after 1797 were subtitled *Militärische Denkwürdigkeiten unserer Zeiten* and numbered separately.

[11] Lieut. G. Scharnhorst, *Handbuch für Offiziere in den angewandten Theilen der Krieges Wissenschaften* (3 vols., Hanover, 1787–90); about 170 subscribers are listed at the beginning of the first vol.

[12] Capt. G. Scharnhorst, *Militairisches Taschenbuch zum Gebrauch im Felde* (Hanover, 1792).

[13] No complete edition of Scharnhorst's works exists. Many unpublished works, some of which are now lost, were printed, however, by his biographers, and large extracts from his major published works were reprinted in C. von de Goltz (ed.), *Militärische Schriften von Scharnhorst* (Berlin, 1881). A new collection is U. von Gersdoff (ed.), *Ausgewählte Schriften*, (Osnabrück, 1983).

Aufklärers, of the importance and value of military knowledge, which were allegedly recognized and expressed by the great generals of history. Passages from several authorities from Folard to Maizeroy and Frederick the Great are cited to drive this point home. The introduction also contains a survey of military literature recommended for the study of the various branches of war. As a basis, Scharnhorst suggests the works of Nicolai and Zanthier. Then, detailed bibliographies are offered for the necessary auxiliary disciplines and the war sciences themselves. In the spheres of tactics, operational activity, and strategy (the new concept is borrowed from Maizeroy; see *Handbuch*, iii. 1–2), the central place is occupied by the works of Maizeroy, Guibert, Turpin, Puységur, Feuquières, Montecuccoli, Folard, de Saxe, Santa-Cruze, and Frederick the Great.[14]

The emphasis on the necessity and usefulness of military theory is also the theme of a work written around 1790, reflecting the developments in military education in Germany and entitled 'On the Utility and Establishment of Military Schools for Young Officers'. Echoing Nicolai, Scharnhorst wrote that a sound theory based on rules and principles explained the successes of Frederick the Great, Gustavus Adolphus, Condé, Caesar, and Alexander. If years of service were sufficient training, old corporals would make generals.[15]

What then is the nature of military theory, and what exactly does it teach? Scharnhorst began to address himself to this question in his early works, reaching his final conclusion by the end of the 1780s. It can be summarized as follows: through conceptualization, military theory makes possible the intellectual treatment of the factors active in war. In his introduction to the *Militair Bibliothek* in 1782, the young Scharnhorst formulated this into a characteristic theoretical framework that accompanied him throughout his life: military theory provided 'correct concepts' (*richtige Begriffe*). These concepts, he wrote three years later in his introduction to the *Bibliothek für Offiziere*, had to be grounded in 'the nature of things or in experience'.

This line of thought is developed in the *Handbuch für Offiziere* in 1787. An inherent interdependence exists between theory and

[14] *Militair Bibliothek* i. 1–38; for a similar survey and a list of the periodical's subscribers, see the introduction to the 2nd issue (1783).

[15] 'Ueber den Nutzen und die Etablirung einer Militär-Schule für die jüngern Offiziere', quoted in Lehmann, *Scharnhorst*, i. 43.

reality. First, one needs clear concepts and principles which clarify the links between the parts of war and the whole; these concepts and principles are necessarily based on the nature of things, and there is no knowledge without them. Then one must understand the actual operation of these concepts and principles in action, for reason alone is not sufficient for developing reality. The application of the concepts and principles to reality requires judgement, which is in turn sharpened only by experience and constant exercise, the major means of which is historical study. Thus, the proper method for educating young officers is, first, to provide them with 'correct theory' and encourage them to think independently and 'clarify their concepts'. This would create a sound basis for analysing experience.[16]

While quite in harmony with the theoretical outlook of the Enlightenment, this theoretical framework reveals a distinctive note and points to several intellectual influences. Firstly, the unique focus on the role of conceptualization in the creation of theory, the relationship between theory and reality, and the link between the parts of war and the whole, is strikingly similar to Montecuccoli's intellectual structure in the introduction to his celebrated *War against the Turks in Hungary*. Indeed Scharnhorst's close affinity to Montecuccoli has been pointed out by Stadelmann. In a letter to a friend written in 1810, Scharnhorst recommended Montecuccoli's work, calling it *Lebensbuch*, and asserted that it had been his constant companion accompanying him through good and bad times.[17]

Scharnhorst's insistence on the insufficiency of reason alone for developing reality also suggests that this theoretical structure and his theoretical interests may have been reinforced by Kant's theory of knowledge and emphasis on the interpretive role of concepts and interdependence of mind and experience. Though no direct evidence for his familiarity with Kant's work is known, the fact that Scharnhorst's early works appeared in 1782–7 makes such an influence very plausible.[18]

Finally, from the 1780s and throughout his life, Scharnhorst saw theory as 'necessarily' grounded not only in 'experience' but also in

[16] *Handbuch für Offiziere*, vol. i, pp. v–vii and 1–4.

[17] Letter of 30 Aug. 1810, in K. Linnebach (ed.), *Scharnhorsts Briefe* (Munich and Leipzig 1914), 404–5; Stadelmann, *Scharnhorst*, pp. 92–9.

[18] Following a general remark by Lehmann, Willhelm Wagner argued for a Kantian influence on Scharnhorst over the issue of the standing armies: W. Wagner, *Die preussischen Reformer und die zeitgenössische Philosophie* (Cologne, 1956), 127–8; in view of the extensive debate on that issue, this argument, like some of Wagner's other conclusions, appears to be rather hasty.

the 'nature of things' this was the characteristic conception which he bequeathed to Clausewitz. Unaware of the part Scharnhorst played in its transference, Raymond Aron, in his treatment of Clausewitz, was the first to call attention to the striking affinity of this conception to Montesquieu.[19] It clearly resembles Montesquieu's famous conception of laws, defined at the opening of the *Spirit of the Laws*, as the 'necessary relations arising from the nature of things'. Indeed, it was revealed by Stadelmann that Scharnhorst ordered the *Spirit of the Laws* from his bookseller in the mid-1790s, which does not exclude an earlier acquaintance with it.[20]

Montesquieu's influence and those of other authors with whom we know Scharnhorst was familiar[21] may also have had much to do with another major feature of Scharnhorst's theoretical approach. As pointed out by Stadelmann, Scharnhorst operated in the midst of a transformation in historical outlook which went hand in hand with a growing sensitivity to the many facets of reality and the interdependence between its component parts.[22] All the military thinkers of the Enlightenment emphasized the paramount value of historical experience. Scharnhorst's works were characterized, however, by a distinctive tendency towards a detailed, concrete, and comprehensive reconstruction of the historical cases in point. Military historians, Clausewitz wrote in his booklet of instructions for the Prussian crown prince,

> invent history instead of writing it . . . The detailed knowledge of a few individual engagements is more useful than the general knowledge of a great many campaigns . . . An example of such an account, which cannot be surpassed, is the description of the defense of Menin in 1794, in the memoirs of General von Scharnhorst. This narrative . . . gives Your Royal Highness an example of how to write military history.[23]

[19] R. Aron, *Clausewitz, den Krieg denken* (Frankfurt am Main, 1980), esp. pp. 163, 308, 331–5.

[20] Stadelmann, *Scharnhorst*, pp. 105–8.

[21] Voltaire's *Siècle de Louis XIV*, Helvétius's *De l'esprit* and *De l'homme*, Rousseau's *Du contrat social*, the writings of Ferguson, Gibbon, and apparently also Möser and Herder were known to Scharnhorst: see Stadelmann, *Scharnhorst*, pp. 102–17; some of this was already known from the biographies of Klippel and Lehmann and from Scharnhorst's letters.

[22] Stadelmann, *Scharnhorst*, p. 119.

[23] Carl von Clausewitz, *Principles of War*, (Harrisburg, 1942), 68–9; see also id., *On War*, II, 6, p. 170. *Die Verteidigung der Stadt Menin* appeared in the *Neues Militärisches Journal*, XI (1803), and in book form in Hanover the same year; reprinted in Goltz (ed.), *Schriften*, pp. 1–58.

Only a detailed historical account can come close to reconstructing the living reality of war, thus achieving some of the value of firsthand experience and conveying the complexity of factors and forces active in war, which may never be explained by a single factor or principle alone.

The 'Development of the General Reasons for the French Success in the Wars of the Revolution', written in 1797 by Scharnhorst and his friend Friedrich von der Decker, provides another analytical example of the same approach. The argument that the French success cannot be reduced to a single factor is the theme of the first chapter, followed by twelve chapters in which the variety of conditions that affected the struggle between the French Revolutionary armies and those of the Allies are traced and presented. These include the political background of the war, the strategic situation of the belligerents, their positions and geographical location, numerical strength and sources of reinforcement and supply, the military organization and methods of warfare, the power of motives, and last but not least, the difference in social infrastructure between the powers of the *ancien régime* and Revolutionary France.[24]

The concrete and comprehensive theoretical approach that characterized Scharnhorst's work from the outset also found typical expression in the definition of the aims of the *Militärische Gesellschaft* that he founded in Berlin in 1801–2. The discussions of the society, according to the first article of regulations, would try to avoid 'one-sidedness' and 'would put theory and practice in proper relationship'.[25]

What then, was Scharnhorst's exact place in relation to the military school of the Enlightenment, and what was his attitude towards it? As mentioned earlier, several factors have contributed to a misrepresentation of Scharnhorst's position on these matters. Scharnhorst was one of the most notable and best-known military *Aufklärers*. Together with his contemporaries, he believed that war was susceptible to intellectual study, theoretical and historical, based upon clear concepts and principles derived from experience. Some branches of war, such as artillery, fortifications, and siegecraft were even

[24] Scharnhorst, 'Entwicklung der allgemeinen Ursachen des Glücks der Franzosen in dem Revolutionskriege', *Neues Militärisches Journal*, VIII (1797); reprinted in Goltz (ed.), *Schriften*, pp. 192–242.

[25] The regulations are cited in Klippel, *Leben* ii. 255–62.

susceptible to a geometrical-mathematical formulation. Hence the supreme importance of the officers' military education and the effort to develop suitable programmes and institutions for this purpose.

Now, as we have seen, within the military school of the Enlightenment there were already gleams of more radical ideas and aspiration, which were unacceptable to Scharnhorst. In his critique in the *Neues Militärisches Journal*, I (1788), of Franz Miller's *Reine Taktik der Infanterie, Cavallerie und Artillerie* (1787–8), Scharnhorst rejected Miller's geometrical and even trigonometrical considerations for battle formation and deployment. Though he believed in the paramount importance of mathematics in the field of fortifications and artillery, as well as in training the officer's mind for logical thinking, he maintained that mathematics could not be applied to the conduct of operations.[26]

The 1790s saw the publication of the works of Lloyd and Tempelhoff and the advent of new trends. Much as Scharnhorst regarded Tempelhoff as a first-rate artillery expert and military historian,[27] he rejected his artificial constructions for the conduct of operations. In a later work which was written in 1811, but which undoubtedly expressed his earlier attitudes, and which clearly betrays the origins of Clausewitz's ideas, he recalled:

> Tempelhoff wrote an essay in which — starting from an arbitrary number of bread and supply wagons — he catalogued all movements that in his opinion an army could undertake. He took supply as the centripetal and operations as the centrifugal force; they balanced at a radius of fifteen miles. This pretty equation made people forget a thousand contradictory experiences. The disease was so catching that the soundest heads were affected.[28]

The novel trends, developing within the legacy of the Enlightenment and relating to the new interest in the conduct of operations, were indeed becoming increasingly influential. Simultaneously, Berenhorst represented a comprehensive reaction against the theoretical tenets of the Enlightenment. Thus, Scharnhorst was now fighting on two fronts. In a critique in the *Neues Militärisches Journal* of Berenhorst's *Nothwendige Randglossen*, Scharnhorst emphasized the advantages of the standing army against Berenhorst's attacks, and also rejected his ironic challenge to the classical conceptions of the military

[26] Also see *Handbuch für Offiziere*, vol. iii, p. v.
[27] See e.g. ibid. i. 4.
[28] Scharnhorst, 'On Infantry Tactics' printed in Paret, *Yorck*, app., 259.

thinkers of the Enlightenment. Where Berenhorst wrote that 'the Prussians won in spite of the art' (*Die Preussen siegten der Kunst zum Hohn*), Scharnhorst replied that 'They won to the honour of the art' (*Sie siegten der Kunst zu Ehren*). While admitting that in the situation they were in, theoretical considerations appeared to be against the Prussians, Scharnhorst argued that, on the other hand, they only won because of their superior organization, discipline, and tactics. Against Berenhorst's undermining criticism, Scharnhorst restated the classical conceptual framework of the Enlightenment: the art of war, like painting and the rest of the arts, has two parts: the one is mechanical and susceptible to theoretical study, the other circumstantial and dominated by creative genius and experience.[29]

Unfortunately for the understanding of Scharnhorst's position in relation to the legacy of the Enlightenment, the last and best-known period of his life, at the outset of the nineteenth century, was also marked by the flourishing of systems for the conduct of operations. These were regarded by him as artificial and one-sided, and stood in contrast to his traditional understanding of the theoretical ideal of the Enlightenment. Against them Scharnhorst directed the main thrust of his criticism in the years in which Clausewitz became acquainted with him and absorbed the fundamentals of his theoretical approach. Clausewitz therefore presented and praised him as an opponent and critic of contemporary military theory, represented by the systems and principles of Bülow, Mathieu Dumas (a well-known historian of the wars of the Revolution who emphasized the key role of high, commanding positions), and Jomini.[30]

To remove any doubt that Scharnhorst did not, in the last period of his life, move away from the position he had held since his youth in the 1780s, but rather that it was the theoretical legacy of the Enlightenment that, so to speak, moved away from him, it is enough to examine his essay 'The Use of Military History, the Causes of its Deficiencies', written in 1806. Here all the themes we have already met are repeated, and the exact scope of Scharnhorst's objection to the new theoretical trends may be seen. The great generals of history, writes

[29] *Neues Militärisches Journal*, XII (1804), 344 ff. For a full reiteration of Scharnhorst's theoretical outlook, made in the same year, see the opening chs. of *Handbuch der Artillerie* (Hanover, 1804), reprinted in *Ausgewählte Schriften*, pp. 153–62.

[30] Clausewitz, 'Über das Leben und den Charakter von Scharnhorst', in L. von Ranke (ed.), *Historisch-Politische Zeitschrift*, I (1832), 197–8.

Scharnhorst—Hannibal, Scipio, Caesar, Turenne, Montecuccoli, and Frederick—studied the principles of the art of war. Some branches of this art are even susceptible to mathematical formulation, but others are dependent on circumstances and cannot be mechanically studied. That is why study alone without genius will never make a great general. One of the branches that has remained without a systematic theory is the conduct of war. In modern times some men, especially the French, have attempted to formulate universal principles for this field, but these have been invalidated by reality and changing experience. Instead, it would be better to concentrate on the study of history. In the education of young officers it leads back to the fundamental rules and principles, and guarantees that the theory of war in all its parts is based on the 'nature of things' and 'experience'.[31]

Far from being the opponent of the traditional conception of military theory, Scharnhorst, in accordance with his general world-view and position throughout his career, was therefore one of the most prominent exponents of the enlightened school of military thought, defending it against reactionary tendencies on the one hand, and against later radical trends which were taking control over it on the other.

Scharnhorst's influence on the young Clausewitz, his pupil and closest protégé, cannot be exaggerated. His role in moulding Clausewitz's political, social, and military views, not to mention the course of his life, was decisive, and his theoretical notions became the basis for Clausewitz's own developing theoretical outlook.

The changing theoretical background against which the two operated, should, however, be stressed first. The generation that separated them gave a totally different starting-point to their thought and theoretical work. The young Clausewitz began his theoretical involvement at the beginning of the nineteenth century, when the theoretical outlook of the Enlightenment was already established and new developments within it created sensation and controversy. To this were added the emergence of a new cultural paradigm and the Napoleonic revolution in warfare. Synthesizing all these trends, Clausewitz emerged as an opponent of what by now had become traditional military theory.

[31] Scharnhorst, 'Nutzen der militärischen Geschichte, Ursach ihres Mangels' (1806), printed in *Ausgewählte Schriften*, pp. 199–207.

Returning to Scharnhorst's legacy; in his youth Clausewitz was attracted, as he was to admit later, by the seductive promise of Bülow's system.[32] These very early notions disappeared entirely when he entered the Berlin Institute for Young Officers. Under Scharnhorst's influence, he—like other disciples of Scharnhorst—rejected the new systems for the conduct of operations as one-sided abstractions which created an intolerable gulf between theory and reality. Instead, he learnt from Scharnhorst that theory had to be concrete and circumstantial, encompass the complexity of political, human, and military conditions that formed reality, and be closely linked to historical experience. Such theory would form free, undogmatic principles, such as Scharnhorst had formulated in his *Handbuch für Offiziere*, and would deal with 'actual war' as Scharnhorst had taught in the Berlin Institute, in contrast to the popular abstractions of the time.[33]

In addition to all this, Scharnhorst also bequeathed to Clausewitz another key conception: theory had to reflect the relationship between the parts of war and the whole, and be 'necessarily grounded in the nature of things'. In essence, there was implicit here a far-reaching theoretical ideal, which was to play a decisive role in Clausewitz's thought.

After the Napoleonic Wars when Clausewitz began to immerse himself in his great theoretical work, he had to clarify for himself, develop, and elaborate the crude, half-intuitive theoretical framework which he had inherited from Scharnhorst, and which he himself had started to work on in his youth.

[32] Clausewitz, 'Bülow', in Hahlweg (ed.), *Verstreute kleine Schriften*, p. 87.
[33] Clausewitz, 'Leben und Charakter von Scharnhorst', pp. 198, 177.

II REFORMULATING MILITARY THEORY IN TERMS OF A NEW INTELLECTUAL PARADIGM

In turning from the military thinkers of the Enlightenment to the study of Clausewitz, a marked difference in the scope, depth, and nature of the treatment accorded to these subjects is clearly noticeable. The military thinkers of the Enlightenment have largely been neglected, their background and collective ideal have not been recognized, and their ideas have been subjected to the polemic and stereotyped criticism which reflect Clausewitz's point of view and the legacy of the German military school of the nineteenth century. Conversely, the domination of this school over the field of military theory secured the 'canonization' of Clausewitz in the late nineteenth and early twentieth centuries, albeit with a somewhat popular and selective interpretation of his thought.

Unfortunately, this imbalance has only been exacerbated in our times. As mentioned earlier, the practical military value of Jomini's work, which had kept the theoretical conceptions of the Enlightenment very much alive, declined sharply after the First World War. By contrast, the interest in Clausewitz, after an eclipse between the two World Wars (except in Germany), was revived in the 1950s, predominantly owing to the significance that his treatment of the relationship between policy and war and of limited war bore on the political and military problems of the nuclear age. A 'Clausewitz renaissance' has developed in strategic and political literature, perhaps no less popular and selective in nature than the attitudes to Clausewitz in the nineteenth century, though, ironically, with opposing emphases.

The rapidly growing involvement of academic historical research has not altered these tendencies either, but, on the contrary, has reinforced them. The main problem has been that the cultural context of Clausewitz's ideas—the transition from the Enlightenment to the German Movement that was hostile to it—has not on the whole been recognized. Indeed, this may already be seen in the confusion that prevails regarding the philosophical influences on his work. The liberation of modern historical study—heralded by Cassirer—from the polemical attitudes that, in the nineteenth century, characterized the campaign of the German Movement against the ideas of the Enlightenment which were labelled as superficial, artificial, and unhistorical, has not reached the military field. In the study of the Enlightenment as a whole, it has been widely recognized that though the accusations of the men of the German Movement had some validity, their hostile fervour drove them into committing against the Enlightenment the very offence with which they had charged it: unsympathetic interpretations that were alien to the values, views, interests, and aims of the period itself. Yet, in the military field, historians and commentators have unwittingly continued to express what was in fact Counter-Enlightenment rhetoric.

For all that, our knowledge and understanding of Clausewitz have been vastly increased since the systematic and academic study of his work began. The works of Hans Rothfels in the 1920s, Herbert Rosinski, Eberhard Kessel, and Walter Malmsten Schering in the three subsequent decades, and Werner Hahlweg from the 1950s, brought to light many of Clausewitz's early writings which are of vital importance to the understanding of his development. To these should be added the critical editions of *On War* published by Hahlweg since 1952. The stages in the development of Clausewitz's work, his military and theoretical ideas, and his political outlook have all received scholarly attention.[1] From the 1950s, the Clausewitz renaissance in strategic and political literature has been matched by an increase in historical studies of Clausewitz, expanding beyond the frontiers of Germany and culminating in the works of Peter Paret and Raymond Aron, both published in 1976.

A few opening remarks on these two books will help to clarify the guide-lines of this work in the study of Clausewitz. Paret's biography *Clausewitz and the State* is the best of its kind, combining extensive research, a remarkable reconstruction of Clausewitz's historical environment, and a sympathetic psychological portrait. Paret also devotes much attention to Clausewitz's intellectual background, and brings together a great deal of relevant material to which the present study is greatly in debt. However, it is the contention of this work that Paret does not fully succeed in placing Clausewitz in his actual intellectual context nor in identifying some of the major influences on his work. He also fails to recognize Clausewitz's theoretical development, particularly the crucial significance and scope of the transformation that took place in 1827

[1] For the authors and works, see throughout my discussion of Clausewitz's ideas. I have taken the same liberty of postponing documentation all through these introductory remarks.

in his way of thinking. As this is coupled with a subtle, unintentional projection of today's attitudes on Clausewitz's thought, Paret also totally misinterprets the essence of Clausewitz's military teaching throughout his life.

Aron's attraction to Clausewitz is especially of interest. Already in the 1950s, he had discovered in Clausewitz a thinker whose ideas closely corresponded to his own regarding the nature of theory in the study of international relations—a problem that had pre-occupied him ever since the outbreak in the early 1950s of the great methodological debate in that field. The far-reaching affinity in their views is revealed in Aron's fundamental political 'realism'; in his rejection of the wider aspirations of the 'scientific school' in the study of international relations; in his rejection of any theory based on a single isolated factor, rendering it artificial and one-sided; in his emphasis on the primacy of historical experience in shaping theory; and last but not least, in his belief that, for all that, the concept of 'theory' can still be given much meaning and possess great value.[2]

From this unique viewpoint Aron offers the most comprehensive and elaborate analysis of Clausewitz's work and theoretical conceptions. The scope of his study is remarkable, and much of his interpretation is penetrating.[3] However, his special affinity to Clausewitz is overshadowed by a serious handicap. Like many of his predecessors, Aron is hardly aware of the cultural context in which Clausewitz worked nor of the intellectual trends to which he gave expression. 'Professing' to a positivist method of interpretation,[4] Aron's theoretical *naïveté* is astonishing. This problem cannot but contribute to the fact that Aron (following in Schering's footsteps, though to a much lesser extent) is inclined to read into Clausewitz's work intellectual patterns and categories which are totally artificial and which obscure even further a subject which is already obscure enough.

All this explains the shift in emphasis and aims in the second part of this study. As mentioned above, the paucity and the largely

[2] For R. Aron's well-known views on these matters, see esp. his *Peace and War, A Theory of International Relations* (New York, 1967), and 'What is a Theory of International Relations?', *Journal of International Affairs*, XXI (1967), 185–206. Also see id., *Clausewitz*, pp. 17–20; since the English edition is substantially abridged, all references are made to the German version.

[3] All references in this work are limited to Aron's first vol. which deals with Clausewitz himself, rather than with his influence in the 19th and 20th cents., which is the subject of Aron's second vol.

[4] Aron, *Clausewitz*, p. 23.

polemic and stereotyped nature of the research on the military thinkers of the eighteenth century has made it necessary to present, in as sympathetic a manner as possible, a general picture of their world-view in the context of their intellectual environment. However, the relative abundance of research on Clausewitz, the prevailing tendencies in viewing his ideas, and, indeed, the intellectual complexity of the subject itself, necessitate a more focused and critical approach from now on. The formation of Clausewitz's conception of theory and criticism of the military thinkers of the Enlightenment will be presented against the background of the new cultural paradigm which emerged in Germany at the turn of the nineteenth century. Then, the development of Clausewitz's efforts to create an adequate military theory of his own will be traced and close attention will be given to the fundamental problems he encountered in the process, which wreaked havoc on his lifelong theoretical outlook and forced him to adopt new ideas and theoretical devices.

Carl Philip Gottlieb von Clausewitz was born in 1780 to a family whose claim to nobility was dubious. His father, who joined the Prussian army when it was in desperate need for men during the Seven Years War, rose to the rank of lieutenant only to be discharged after the war when Frederick purged the Prussian officercorps of middle-class elements. After Frederick's death, he succeeded, however, in securing appointments as NCOs for three of his sons. The twelve-year-old Carl began his military service in an infantry regiment in 1792, and in 1793–5 he took part in the campaigns of the First Coalition against Revolutionary France. The following six years of peace were spent by the young lieutenant in the provincial garrison town of Neuruppin. He left it only in 1801 when he was admitted into the Institute for Young Officers in Berlin, which had been revived, enlarged, and thoroughly reformed by Scharnhorst, who had shortly before entered the Prussian service. This was a turning-point in Clausewitz's life. During his three years of study at the Institute he made the acquaintance of Scharnhorst, absorbed the foundations of his military outlook, and became his closest protégé. His education was broadened dramatically, and new intellectual horizons were opened. After finishing first in his class, he was on the road leading to the centre of the political and military events in the Prussia of the Napoleonic Wars, of reform, and of the Restoration.

In 1804 Clausewitz was appointed adjutant to Prince August, cousin of Frederick Wilhelm III King of Prussia. In this capacity and as a brevet captain, he took part in the Battle of Auerstädt (1806), and after Prussia's catastrophic defeat, he and the Prince fell into French captivity. At the end of 1807 the two returned from their imprisonment in France, and at the beginning of 1809 Clausewitz was co-opted by Scharnhorst as his assistant in the *Allgemeine Kriegsdepartement*, the nucleus of a new ministry of war. As head of the department, Scharnhorst orchestrated the military reforms of Prussia, championed and carried out by a group of young officers. Among the reformers, Clausewitz made the acquaintance of Gneisenau, Scharnhorst's major ally, who became an intimate friend. During this period he also married Countess Marie von Brühl, who had been his fiancée for five years. Their uniquely close attachment is revealed in their correspondence, which constitutes the principal source for Clausewitz's biography. It was Marie who published Clausewitz's posthumous works.

Clausewitz's military career was continuously matched by intensive intellectual activity. His strong interest in military theory dates at least from his days at the Institute. His early writings refer, among others, to Montecuccoli, Feuquières, Santa-Cruz, Folard, de Saxe, Puységur, Turpin, Guibert, Frederick, Lloyd, Tempelhoff, Berenhorst, Bülow, Dumas, Venturini, Massenbach, and Jomini.[5] And in a series of works written during his twenties and early thirties, he formulated the theoretical conceptions which were to find their final place in his major work, *On War*.

In 1810 Clausewitz was appointed major in the General Staff, instructor in the new Officers' Academy, and military tutor to the Prussian crown prince. His work during the reform era, motivated by the desire to see Prussia liberated through the destruction of the Napoleonic Empire, culminated in 1812. With the French invasion of Russia, Clausewitz, like some of his comrades, left Prussia and joined the Russian army, acting against the instructions and policy of his king. In Russia he was promoted to colonel, served in various staff posts, and took part in the Battle of Borodino.

After Napoleon's retreat and despite the fact that Prussia joined the war against France, Frederick William III refused to accept

[5] Hans Rothfels, *Carl von Clausewitz, Politik und Krieg* (Berlin, 1920), 29–30; Paret, *Clausewitz*, p. 81.

Clausewitz back into the Prussian service. His friends, however, arranged for him to be attached to Blücher's headquarters as a Russian liaison officer, and again working together with Scharnhorst, Clausewitz played an important role in the Prussian command at the battles of Bautzen and Lützen (Scharnhorst was mortally wounded during the latter). Since all efforts to obtain the king's pardon failed, Clausewitz was compelled to serve in the German Legion of volunteers and in secondary theatres of operations for the duration of the campaigns of autumn 1813 and of 1814. Only after Napoleon's defeat was he accepted back into the Prussian service. In the campaign of 1815 he served as the chief of staff to the corps which contained Grouchy at Wavre, while the main body of the Prussian army marched to join Wellington at Waterloo.

After the war, Clausewitz was appointed chief of staff to the force stationed in Prussia's newly acquired territories along the Rhine, and he remained at Koblenz in that capacity until 1818. He was then promoted to general and appointed head of the Military Academy at Berlin, largely an administrative function. The end of the era of war and the beginning of a long period of peace paralleled the triumph of the Restoration in Prussia. The disappearance of the external challenge of his youth and the king's suspicious attitude towards his radical reputation, which clouded his military career, made Clausewitz concentrate on the intellectual interests which had hitherto been overshadowed by his military activities. During his time at Koblenz, Clausewitz made the first attempt to write a general theoretical work on war, and this was followed by a continuous period of work while serving in Berlin. In 1830, the course of the work was interrupted by Clausewitz's appointment as commander of one of the artillery divisions of the Prussian army. A short time later, with the outbreak of the revolutions of 1830, he was appointed chief of staff to the army raised under Gneisenau in anticipation of possible Prussian intervention in Poland. In 1831, both men fell victim to the great cholera epidemic which swept across the continent.[6]

[6] This biographical sketch is merely intended to provide a framework for the study of Clausewitz's intellectual development. The first biography of Clausewitz, incorporating his letters and some of his unpublished works, was written by Karl Schwartz, *Leben des Generals von Clausewitz und der Frau Marie von Clausewitz* (2 vols.; Berlin 1878); amendments and supplements, particularly regarding Clausewitz's family and childhood, were introduced by Eberhard Kessel, 'Carl von Clausewitz: Herkunft und Persönlichkeit', *Wissen und Wehr*, XVIII (1937); for the recent and by far the best biography, see Paret, *Clausewitz*.

In his early twenties Clausewitz absorbed Scharnhorst's criticism of the new systems of operations as one-sided abstractions, divorced from reality. Simultaneously, Clausewitz's intellectual environment powerfully projected the message that the world-view of the French Enlightenment, on which the old theory of war was based, was fundamentally false. Since the 'Storm and Stress' period, the ideas of the French Enlightenment had been labelled artificial, superficial, and pretentious. And this became the prevailing cultural and political outlook in Germany at the advent of the nineteenth century following the disillusion with the French Revolution and the fierce reaction against Napoleonic imperialism.

A classic example of the outlook and sentiment of the time can be found in the comparison Clausewitz drew in late 1807, on his return from French captivity, between the national characteristics of the French and the Germans. French feelings and thinking, he wrote, were active, excited, and quick, but also shallow and always prepared to sacrifice content for form and appearance. By contrast, German feelings and thinking were calm, deep, and penetrating, and they strove toward comprehensive expression and understanding.[7] That Clausewitz was here expressing prevailing ideas propounded for example by Möser, Wilhelm von Humboldt, the Romantics, and Fichte, has already been noted by some of Clausewitz's interpreters.[8] In another, later, classical example of contemporary attitudes in Germany, Clausewitz criticizes the views of 'philosophers who are right about everything by means of universal concepts', being 'strongly influenced by Parisian philosophy and politics'.[9]

Clausewitz's cultural environment was not only critical of the legacy of the Enlightenment but also provided him with an alternative conception of reality, to be used as a basis for a reformulation of military theory. Berenhorst had already given expression to some of the most distinctive themes of the new climate of ideas. The young Clausewitz now developed a different, more comprehensive, and sophisticated synthesis of the new intellectual themes, stressing the diversity and living nature of human reality and centring on the conceptions of rules, genius, moral forces, factors of uncertainty, and history.

[7] 'Die Deutschen und die Franzosen', in Hans Rothfels (ed.), *Carl von Clausewitz, Politische Schriften und Briefe* (Munich, 1922), esp. pp. 37–45.

[8] Rothfels, *Clausewitz, Politik und Krieg*, pp. 113–16; Paret, *Clausewitz*, pp. 133–4.

[9] 'Umtriebe', in Rothfels (ed.), *Schriften*, p. 166.

We have seen that the military thinkers of the Enlightenment drew their conception of theory, based on the twin concepts of rules and genius, from the legacy and development in the Enlightenment of the seventeenth-century neo-classical theory of art. Into this theory there were injected, throughout the eighteenth century, increasing emphases on the role of free, creative genius, a development which was also reflected in the works of the military thinkers of the Enlightenment. With Kant the transformation in the eighteenth-century outlook on the theory of art was completed, and the emphases were finally reversed. Genius did not embody the rules as had been believed by the neo-classicists. Nor was it an essential, creative, and imaginative force, as important as the rules themselves. Genius was rather the exclusive source of all artistic creation which could not be adequately formalized in any set of rules. It was the measurement of all rules which were only justified as crude means for capturing, by way of concepts, something of its creative force. 'Genius', wrote Kant in his *Critique of Judgement*, 'is the talent (natural endowment) which gives the rule to art . . . [it] is a talent for producing that for which no definite rule can be given.' The genius's example can merely provide 'a methodical instruction according to rules, collected, so far as the circumstances admit.'[10]

The fact that Clausewitz's conception of military theory was rooted in Kant's theory of art was for the first time and most clearly pointed out in 1883 by Kant's student, the philosopher Hermann Cohen, and has since been repeated by all of Clausewitz's major interpreters (in contrast to much uninformed comment chiefly by non-German authors).[11] Although no direct evidence as to Clausewitz's familiarity

[10] Immanuel Kant, *The Critique of Judgement* (Oxford, 1961), esp. articles 46–50; the quotations are from pp. 168, 181.

[11] See esp. Hermann Cohen, *Von Kants Einfluss auf die deutsche Kultur* (Berlin, 1883), 31–2; Rothfels, *Clausewitz, Politik und Krieg*, pp. 23–5; Walter Malmsten Schering, *Die Kriegsphilosophie von Clausewitz* (Hamburg, 1935), 105–11, and id., *Wehrphilosophie* (Leipzig, 1939), 343–4; Erich Weniger, 'Philosophie und Bildung im Denken von Clausewitz', in W. Hubatsch (ed.), *Schicksalswege Deutscher Vergangenheit* (Düsseldorf, 1950), 123–43; Paret, *Clausewitz*, esp. pp. 160–3; and Werner Hahlweg, esp. 'Philosophie und Theorie bei Clausewitz', in Clausewitz Gesellschaft (ed.), *Freiheit ohne Krieg* (Bonn, 1980), 325–32. Schering was the first to argue that Clausewitz may have also been influenced by 18th-cent. German aesthetical thinkers, such as Sulzer and Lessing, who paved the way for Kant (*Wehrphilosophie*, p. 343). While this may obviously be true and applies to the whole break from Gottsched's neo-classicism pioneered by Bodmer and Breitinger, Clausewitz's conceptions are clearly Kantian and whether he was familiar with Kant's predecessors is purely conjectural.

with Kant's works exists, we know that Clausewitz was introduced to them through the lectures of Kiesewetter, one of Kant's best-known popularizers and one of the pillars of the Institute for Young Officers where he was the instructor of mathematics and logic.[12]

Like the military thinkers of the Enlightenment, and even more consciously than them, Clausewitz found in the theory of art a highly suggestive model for the theory of the 'art' of war. Both dealt with the theory of action; in both, given means were employed to achieve a required effect through a creative process which involved principles of an operational nature. From his earliest works to *On War*, Clausewitz adapted Kant's theory of art to criticize the work of the military thinkers of the Enlightenment, and to develop his own conception of the theory of war.

Already in 1805, Clausewitz had employed the new conceptual framework in his criticism of Bülow. Bülow's definitions of strategy and tactics, he argued, were invalid, because Bülow did not state their purpose. Stating the purpose is essential to the definition of art which is 'the use of given means to achieve a higher end'.[13] Furthermore, Clausewitz objected to Bülow's opinion that, if need be, the general ought to follow his genius above and contrary to the rules:

> one *never* rises above the rules, and thus when one appears to go *against* a rule, one is either *wrong*, or *the case does not fall under the rule any more* . . . he who possesses genius ought to make use of it, *this is completely according to the rule!*[14]

Any division or conflict between genius and rules was now inadmissable. In a fragment written in 1808 or 1809 Clausewitz reasserted this: 'genius, dear sirs, never acts contrary to the rules'.[15]

In an essay 'On Art and Theory of Art'—written at an unknown time, perhaps after Clausewitz's period of study in the Institute but possibly only in the late 1810s or early 1820s as a preparatory work for the writing of *On War*—the conception of theory is elaborated upon as Clausewitz strives to clarify his ideas. Like Kant, he distinguishes between science, whose aim is knowledge through

[12] Some of Clausewitz's notes, taken in one of Kiesewetter's lectures on mathematics, were found by Schering in Clausewitz's family archive (now lost); Schering, *Kriegsphilosophie*, pp. 105 ff.

[13] 'Bülow', in Hahlweg (ed.), *Verstreute kleine Schriften*, pp. 67–8.

[14] Ibid. 80–1.

[15] The fragment, 'Tactische Rhapsodien' was never printed and appears to have been lost. The quotation is from Rothfels, *Clausewitz, Politik und Krieg*, p. 156.

conceptualization, and art, whose essence is the attainment of a certain aim through the creative ability of combining given means. Between the two concepts there exists, Clausewitz points out, a certain overlapping, and art is assisted by knowledge. Thus, 'the theory of art teaches this combination [of means to an end] as far as concepts can . . . Theory is the representation of art by way of concepts.'[16] However, this representation is fundamentally very limited and varying. In his notes on strategy (1809), Clausewitz wrote for example, following Scharnhorst, that 'the part of strategy that deals with the combination of battles must always remain in the sphere of free (unsystematic) reasoning'.[17]

All these themes receive comprehensive treatment in Book II of *On War*, 'On the Theory of War'. Clausewitz again presents the distinction between a science of concepts and an art of creative capability. War fits much more into the model of art, while the title science is better kept for fields such as mathematics and astronomy. However, Clausewitz also makes it clear that these are no more than analogies. The major difference between the nature of creative activity in the arts and in war is that in war the object reacts. From this point of view, as well as from that of its subject-matter, war belongs much more to the field of social intercourse, being close to commerce and above all to politics.[18]

The various systems for the conduct of operations are again accused of being abstracted from reality and separating genius from rules:

> Anything that could not be reached by the meagre wisdom of such one-sided points of view was held to be beyond scientific control: it lay in the realm of genius which *rises above all rules*. Pity the soldier who is supposed to crawl among these scraps of rules, not good enough for genius, which genius can ignore, or laugh at. No; what genius does is the best rule and theory can do no better than show how and why this should be the case.[19]

[16] The fragment: 'Über Kunst and Kunsttheorie' was printed by W. M. Schering (ed.), *Clausewitz, Geist und Tat* (Stuttgart, 1941); see esp. pp. 154–5, 159. For Kant's distinction between science and art, see his *Critique of Judgement*, article 43, pp. 162–4. Three other fragments on the theory of art were also printed by Schering in *Geist und Tat*.

[17] 'Strategie' (1809), in Hahlweg (ed.), *Verstreute kleine Schriften*, p. 61.

[18] *On War*, II, 3, pp. 148–50.

[19] Ibid. II, 2, p. 136.

Appealing to the genius who is supposed to stand above the rules 'amounts to admitting that rules are not only made for idiots, but are idiotic in themselves.'[20]

The relationship between rules and genius is therefore clearly concluded in terms of the new paradigm in the theory of art:

> It is simply not possible to construct a model for the art of war that can serve as a scaffolding on which the commander can rely for support at any time . . . no matter how versatile the code, the situation will always lead to the consequences we have already alluded to: *talent and genius operate outside the rules, and theory conflicts with practice.*[21]

The emphasis with Clausewitz, therefore, shifts from the rules to the freely creating genius. Genius, however, is not a new sort of abstraction. It is a quality belonging to living men whose activity is dependent on their particular psychological profile, motivations, and aims, as well as on the conditions of their environment. Rejecting dead abstractions for real life and acting personalities was a dominant theme in German cultural outlook and artistic creation since the 'Storm and Stress' period. It remained at the centre of Goethe's and Schiller's outlook in their mature works. And its importance for the Romantics cannot of course be exaggerated. Here too, Clausewitz gave expression to a new world-view whose domination over Germany, when he started his intellectual and literary activities in the first years of the nineteenth century, was already secure.[22] An interesting fact, pointed out by Paret, is that Schiller, the author of historical dramas based on charismatic personalities—(*The Maiden of Orleans*, *William Tell*, *Mary Stuart*, and *Wallenstein* is the author most frequently mentioned in Clausewitz's letters.[23] Schiller is also known as the most philosophically inclined among the great German artists of the late eighteenth century, as Kant's disciple, and as the author of aesthetical works in which he stressed the free operation of genius.[24]

[20] *On War*, III, 3, p. 184.

[21] Ibid. II, 2, p. 140.

[22] Oestreich's suggestion that Clausewitz's conception of genius owed something to the neo-stoical tradition in the early modern period as reflected in the German *Klassizismus* might be, broadly speaking, true, though Oestreich relies on Rothfels's and Schering's very incomplete interpretation of Clausewitz's immediate and dominant intellectual background; Oestreich, *Neostoicism and the Early Modern State*, p. 88.

[23] Paret, *Clausewitz*, p. 84.

[24] For Schiller's aesthetic conceptions and Kant's philosophy, see e.g. R. D. Miller, *Schiller and the Ideal of Freedom, A Study of Schiller's Philosophical Works with Chapters on Kant* (Oxford, 1970).

It is, therefore, not surprising that Clausewitz's emphasis on the role of the creative personality constitutes, as Paret notes here too, one of the striking differences between his outlook and that of Scharnhorst.[25] The explanation for that goes, however, further than Paret's suggestion of variations in interests or aims between the two. This difference offers, in fact, a classic demonstration of the paradigmatic change between the teacher and his pupil. Scharnhorst was a typical representative of the military school of the Enlightenment, which was institutionally and structurally oriented. Characteristically, the military thinkers of the Enlightenment interpreted Frederick's victories chiefly as a product of the Prussian battle deployment. And the legacy of the Enlightenment, adapted by Jomini, continued its reign, interpreting Napoleon's sensational successes in utterly impersonal terms. Neither Frederick nor even Napoleon drew Clausewitz's attention to the role of the great personality; a new world-view was needed for that, and again, it may be traced to his earliest works.

In his notes on strategy of 1804, Clausewitz wrote that a strategic plan 'is a pure expression of [the general's] manner of thinking and feeling, and almost never a course chosen by free consideration'.[26] In this provocative argument he expanded Machiavelli's well-known point and also cited the example the latter used: Fabius *cunctator* 'did not delay operations against the Carthaginians because this type of war so suited circumstances, but rather because it was his nature to delay'.[27]

This point of view, which elevates the general's personality above any abstract strategic considerations, is also strikingly manifest in Clausewitz's interpretation of the operations of Gustavus Adolphus and Frederick the Great. In 'Gustavus Adolphus's Campaigns of 1630–1632', apparently written during the Napoleonic period,[28] Clausewitz presents the personality and motivations of the king and his adversaries as the key to the events of the war—clearly a conscious antithesis to the military thinkers of the Enlightenment.[29] Schiller's famous trilogy *Wallenstein*, published

[25] Paret, *Clausewitz*, p. 166.

[26] 'Strategie' (1804), in Hahlweg (ed.) *Verstreute kleine Schriften*, p. 10.

[27] Ibid.

[28] 'Gustav Adolphs Feldzüge von 1630–1632', *Hinterlassene Werke* (Berlin, 1832–7), vol. ix; for the date of composition see the editor's introd., p. vi.

[29] This was first pointed out by Rothfels, *Clausewitz, Politik und Krieg*, pp. 61–9; touched upon in Kessel's introd. to the first edn. of *Strategie* (Hamburg, 1937), p. 24; and developed in Paret, *Clausewitz*, pp. 85–8.

in 1800, and one of Clausewitz's favourite works, may very well have influenced both Clausewitz's choice of subject and manner of treatment.[30]

Historical study, writes Clausewitz, dwells on 'the mathematical level of physical forces' and ignores the subjective forces in war; yet, it is precisely these forces which are the most decisive.[31] To understand the events of the war, one should understand the particular psychological profile of the operating individuals in the context of their particular milieu. 'Is it not wiser to pay less attention to what the enemy *can* do and pay more attention to what he *will* do? . . . here lies a more fruitful field for strategy than the degrees of angles of operations.'[32]

The idea stressed in 'Gustavus Adolphus' is again sharply expressed in Clausewitz's note on strategy of 1808, directed against Jomini's analysis of the campaigns of Frederick the Great. As we have already noted in the chapter on Jomini, Clausewitz rejected the substitution of abstract, lifeless principles for Frederick's complex and concrete reality and particular psychology:

> To appreciate the value of his [Jomini's] abstractions, one must ask if one wants to give up all of Frederick II's practical life as a general for these couple of general maxims which are so easy to grasp? . . . did Frederick violate these maxims out of ignorance? . . . It is impossible to hang [the diversity of Frederick's generalship] . . . on a couple of meagre ideas . . . What is the conclusion of all this? That the general's temper greatly influences his actions . . . that one must not judge generals by mere reason alone.[33]

Not only was the abstract intellectual interpretation of the activities of great generals deemed to be fundamentally artificial, but so was

[30] Schiller's reputation and career as a historian, which culminated in his appointment as professor of history at Jena, is overshadowed by his dramatic and philosophical achievements. *Wallenstein* was preceded by a widely read *Geschichte des dreissigjährigen Krieges* (1791–3) in which he was already trying to uncover the proper relation between the great personality and the conditions of his time. Also see: W. M. Simon, *Friedrich Schiller, the Poet as Historian* (Keele, 1966); and Lesley Sharpe, *Schiller and the Historical Character* (Oxford, 1982). For Clausewitz's reference to Yorck's inquiry of the troops' mood at the decisive meeting in Tauroggen when he made up his mind to take his corps out of the Napoleonic army, as recalling Schiller's *Wallenstein*, see Clausewitz, *The Campaign of 1812 in Russia* (London, 1843), 239 (*Hinterlassene Werke*, vol. vii); Paret, *Clausewitz*, p. 230.

[31] 'Gustav Adolph', *Hinterlassene Werke*, ix. 8.

[32] Ibid. 46.

[33] 'Strategie' (1808), Hahlweg (ed.), *Verstreute kleine Schriften* pp. 47–9.

the excessive emphasis on intellectual faculties and necessary knowledge. In his notes on strategy of 1804, Clausewitz lists the disciplines that the military thinkers of the Enlightenment carefully compiled for their educational programmes for officers: mathematics, map drawing, geography, artillery, fortifications, siegecraft, entrenchments, tactics, and strategy. Regarding each subject, he concludes that the general only requires a broad but sound, rather than a detailed, knowledge. He has no need for 'professorial' or 'pedantic' knowledge, and can manage with a 'few abstract truths'. What he predominantly requires is sound judgement and a strong character: 'a strong, ambitious spirit'.[34]

Clausewitz's clearly ironic attitude towards the Enlightenment ideal of knowledge is again manifest in his *Principles of War for the Crown Prince* (1812):

> Extensive knowledge and deeper learning are by no means necessary [for the general], nor are extraordinary intellectual faculties . . . For a long time the contrary has been maintained . . . because of the vanity of the authors who have written about it . . . As recently as the Revolutionary War we find many men who proved themselves able military leaders, yes, even military leaders of the first order, without having had any military education. In the case of Condé, Wallenstein, Suvorov, and a multitude of others it is very doubtful whether or not they had the advantage of such education.[35]

The last sentence in particular, which is a straightforward rejection of one of the major doctrines of the Enlightenment, once again demonstrates the paradigmatic shift between Clausewitz and his mentor. It was clearly at variance with Scharnhorst's lifelong beliefs and statements.

Clausewitz discusses the qualities that a general requires in his treatment of military genius in *On War*, which will not be elaborated upon here. The important point is again that character and spirit are more essential than cognitive faculties; fundamentally, war is an activity more than an intellectual discipline. Even the required cognitive qualities are of the empirical and applied sort.[36] It is true that Clausewitz twice repeats Napoleon's dictum that the complexity of the problems involved in war is of the order of mathematical problems that would require a Newton. However what distinguishes

[34] 'Strategie' (1804), Hahlweg (ed.), *Verstreute kleine Schriften*, pp. 6–8.
[35] *Principles of War*, p. 60.
[36] See esp. *On War*, I, 3, 'On Military Genius'.

military knowledge is its relation to life. 'Experience, with its wealth of lessons, will never produce a *Newton* or an *Euler*, but it may well bring forth the higher calculations of a *Condé* or a *Frederick*.'[37]

Clausewitz's emphasis on the general's personality, emotions, and motivations went hand in hand with his emphasis on the decisive role of the moral forces that animate armies. Here too, as we have seen in Jomini's case, the understanding of the change in the intellectual paradigm is essential. The military thinkers of the Enlightenment were far from ignoring the importance of moral forces, and Lloyd even offered an extensive study on the subject, adapting the conceptions and views of the contemporary psychology of desires. However, on the whole, they regarded moral forces as too elusive and belonging to the sublime part of war. And since they were interested in intellectual control, they saw no point in discussing moral forces at length. The intellectual transformation generated by the men of the 'Storm and Stress' period and the Romantics, which placed man's inner world at the centre of human experience, involved a radical change in the interpretation of, and regard for, the ideal of knowledge. The new perspective was largely rooted in anti-rationalistic trends, and thus the focusing on uncontrollable elements was for many of its exponents a special point to be made rather than a sacrifice. The Enlightenment ideal of understanding and control was substituted by a comprehensive and vitalistic one, and consequently the standards for what was considered significant and worth discussing also changed.

Without attempting an impossible summary of the comprehensive intellectual environment and its influences on Clausewitz, it is nevertheless worth noting the following points: that Clausewitz shared with his wife the universal admiration for Goethe and Schiller and in fact, as was probably common with courting couples, *Werther* was a subject of conversation during one of their first meetings;[38] that upon their return from captivity in 1807, Clausewitz and Prince August were the guests of Madame de Staël in her famous place of exile at Coppet in Switzerland for two months, where Clausewitz made the acquaintance of August Wilhelm Schlegel, with whom he

[37] *On War*, II, 2, p. 146; I, 3, p. 112; VII, 3, p. 586.

[38] See Marie's description of her acquaintance with her husband in Schwartz, *Leben*, i. 185.

was impressed despite the fact that he was far from accepting his world-view as a whole;[39] and that prominent Romantic poets and dramatists such as Achim von Arnim, Clemens Brentano, Heinrich von Kleist, and Friedrich, Baron de la Motte Fouqué, as well as Fichte and Schleiermacher, moved in the same social circle in Berlin as the Clausewitzes.[40]

If the origins of Clausewitz's conception of moral forces are wide and varied, its nature is easier to define. Firstly, it is clear that he rejected both idealism and mysticism. 'I recognize', he wrote, 'no pure spiritual thing apart from thoughts; all notions, even all sensations with no exceptions, are a mixture of spiritual and material nature.'[41] Clausewitz's relation to the various themes in Romanticism is strikingly summarized by Peter Paret:

> He benefited enormously from the liberating emphasis that the early Romantics placed on the psychological qualities of the individual; but he did not follow such writers as Novalis or the Schlegel brothers in their surrender to emotion. The religious wave of Romanticism did not touch him; nor did its mysticism, nostalgia, and its sham-medieval, patriarchal view of the state. In feeling and manner he was far closer to the men who had passed through the anti-rationalist revolt of the 'Sturm und Drang' to seek internal and external harmony, and who gave expression to their belief in the unity of all phenomena.[42]

The emphasis on moral elements is already very distinctive in the notes on strategy of 1804, and, as we have seen, it is given systematic expression in the criticism of Bülow and the legacy of the Enlightenment. According to Clausewitz, emotional forces were indeed difficult to determine and control, but they were essential not only for a true, comprehensive, and living conception of war, but also for understanding the nature and boundaries of its theory. In his quest for precision, Bülow concentrates on the material elements which are susceptible to mathematical calculations, and ignores the moral forces that animate war. He thus misrepresents the real nature of war, and creates a mechanistic and one-sided theory.[43]

[39] Clausewitz's letter to his fianceé, 5 Oct. 1807: Schwartz, *Leben*, i. 299.

[40] Hagemann, *Von Berenhorst zu Clausewitz*, p. 69; Paret, *Clausewitz*, p. 212.

[41] From a fragment written in 1807–8, 'Historisch-Politische Aufzeichnungen', in Rothfels (ed.), *Schriften*, p. 59; the metaphysical conception expressed in the passage is, incidentally, clearly Kantian.

[42] Paret, *Clausewitz*, p. 149.

[43] 'Bülow', in *Verstreute kleine Schriften*, pp. 79, 81.

Several statements that Clausewitz made during the reform era reflect the new cultural paradigm in a particularly classical manner. Immediately after rejecting the fantasies of the new mystical sects, Clausewitz goes on to write that they nevertheless express a genuine need of the time, 'the need to return from the tendency to rationalize to the neglected wealth of feeling and fantasy'.[44] On 11 January 1809, in response to an article that Fichte wrote on Machiavelli, Clausewitz sent a letter to the famous philosopher, in which he criticized 'the tendency, particularly in the eighteenth century [to] form the whole into an artificial machine, in which the moral forces were subordinated to the mechanical'. Conversely, he wrote, the 'true spirit of war seems to me to lie in mobilizing the energies of every individual in the army to the greatest possible extent, and in infusing him with bellicose feelings, so that the fire of war spreads to all elements of the army'. That would be the end of the old attitudes, 'for in every art the natural enemy of mannerism is the *spirit*'.[45]

In three separate discussions in *On War*, Clausewitz outlines the moral forces that motivate war, expanding the ideas presented in the critique of Bülow written in 1805.[46] The problem with military thinkers is that 'they direct their inquiry exclusively towards physical quantities, whereas all military action is intertwined with psychological forces and effects'. Thus, 'it is paltry philosophy if in the old fashioned way one lays down rules and principles in total disregard of moral values'.[47]

The one-sided nature of the old theory stems from a genuine difficulty:

> Theory becomes infinitely more difficult as soon as it touches the realm of moral values. Architects and painters know precisely what they are about as long as they deal with material phenomena . . . but when they come to the aesthetics of their work . . . the rules dissolve into nothing but vague ideas.[48]

Moral forces do not evade theoretical treatment altogether. A series of patterns 'in the sphere of mind and spirit have been proved by

[44] Rothfels (ed.), *Schriften*, p. 59.

[45] Schering (ed.), *Geist und Tat*, pp. 77, 78, 80; Paret, *Clausewitz*, pp. 176–7. Compare with W. von Humboldt, p. 243 below.

[46] *On War*, I, 4–5; II, 2, pp. 137–9; III, 3–7.

[47] Ibid. II, 2, p. 136; III, 3, p. 184.

[48] Ibid. II, 2, pp. 136–7.

experience: they recur constantly, and are therefore entitled to receive their due as objective factors'. Yet, in general, moral forces 'will not yield to academic wisdom. They cannot be classified or counted. They have to be seen or felt.'[49]

The effect of moral forces as well as the bilateral nature of war are among the main factors which turn war into a field saturated with the unknown and unforeseen, and create a gulf between planning and the actual course of war. Here too the gap between the military thinkers of the Enlightenment and Clausewitz fits the pattern we have already met. The Enlightenment thinkers were quite aware of the factors of uncertainty but focused on what they considered to be suitable for intellectual formulation. Clausewitz regarded their attitude as dogmatic and divorced from reality, and demanded an all-encompassing theory. 'They aim at fixed values; but in war everything is uncertain, and calculations have to be made with variable quantities.'[50]

It is illuminating to compare this with the works of the Prussian general Friedrich Constantin von Lossau (1767–1848), a participant in Scharnhorst's *Militärische Gesellschaft* and one of the reformers, whose book *War* (1815) elaborated many of the ideas later to become famous in Clausewitz's *On War*. Because of the great progress which had been made in the sciences and the arts in the last centuries, wrote Lossau, people sought similar achievements in the study of war. They forgot, however, the decisive influence of the human personality and of chance in war, to which Berenhorst was the first to call attention.[51]

Clausewitz again expressed the attitudes of his intellectual environment but this time a suitable concept was less at hand. Thus, though he had emphasized the uncertainties involved in war from his early works, he only adopted the concept of 'friction' at a later stage, initially in the *Principles of War for the Crown Prince* of 1812.[52] 'The

[49] Ibid. III, 3, 184.

[50] Ibid. II, 2, p. 136.

[51] F. von Lossau, *Der Krieg* (Leipzig, 1815), 284–8; the book deals extensively with the warrior's intellectual and moral faculties, presenting war as a clash of wills motivated by patriotic and other psychological energies. On Lossau see Hagemann, *Von Berenhorst zu Clausewitz*, pp. 44–55.

[52] The relatively late appearance of the concept of friction has been pointed out by Kessel, 'Zur Genesis der modernen Kriegslehre', *Wehrwissenschaftliche Rundschau*, III/9 (1953), p. 408. Rothfels (*Clausewitz, Politik und Krieg*, p. 90) has called attention to a very similar formulation in *The Spirit of the Laws*, Bk. XVII, ch. 8, where Montesquieu wrote that, like in mechanics, frictions often change the implications of theory. However, the popularity of this mechanistic image in the 18th cent. makes any direct inference pointless.

conduct of war', he wrote, 'resembles the working of an intricate machine with tremendous friction, so that combinations which are easily planned on paper can be executed only with great effort.'[53]

This idea is reiterated in *On War*. The gulf between planning and reality is mainly rooted in the enormous complexity of factors involved, whose effects are difficult to foresee. This is all the more so since war is characterized by the 'uncertainty of all information' which means that 'all action takes place, so to speak, in a kind of twilight'.[54]

Clausewitz's demand for a theory which fully expresses the diversity of reality is closely related to the emergence of a new outlook on history that he introduced into the study of war. His place in the rise of historicism, pointed out by some of his interpreters,[55] is of paramount importance to the understanding of his theoretical outlook and the tensions inherent in it.

As shown by Paret, Clausewitz's early works already contain references to Machiavelli, Montaigne, Montesquieu, Robertson, Johannes von Müller: the historian of the Swiss confederation, Ancillon: the Prussian conservative anti-Enlightenment and anti-Revolutionary historian and statesman, and Gentz: the arch-conservative and disciple of Burke. Evidence for Clausewitz's familiarity with the works of Herder and Möser exists only at later stages of his life but he probably read them much earlier. This historical reading blended with the dominant influence of Scharnhorst's concrete, particularist, and circumstantial approach to the past.[56]

Again one should look at the German intellectual environment in which Clausewitz operated. Moving in the same social circle in Berlin as the Clausewitzes were Adam Müller, Savigny, and Schleiermacher.[57] The first was the most prominent spokesman of

[53] *Principles of War*, pp. 61–8; the quotation is from p. 61.

[54] *On War* I, 7; II, 2, p. 140.

[55] This was well treated by Rothfels, *Clausewitz, Politik und Krieg*, pp. 61–9; noted by Kessel (following the appearance of Meinecke's *Historismus* in 1936) in his introd. to the first edn. of *Strategie*, p. 11; and was lately discussed at length by Paret, *Clausewitz*.

[56] Paret, *Clausewitz*, pp. 81–2. See ibid. 312 for Clausewitz's note to the library in 1820 requesting, among others, a collection of Herder's essays and anthology of Greek lyrics and epigrams, *Herders Zerstreute Blätter*. For a reference to Möser see 'Umtriebe', which is much in affinity with Möser's views, Rothfels (ed.), *Schriften*, p. 164.

[57] Hagemann, *Von Berenhorst zu Clausewitz*, p. 69; Paret, *Clausewitz*, pp. 212, 316.

the historical approach to politics and economics; the second, the founder of the 'historical school' of jurisprudence; and the third, the one who offered a historically conscious explanation for the diversity of religious faith. Rejection of the universal abstractions of the Enlightenment in favour of the belief in historical diversity and the complexity of the forms of society and politics was one of the dominant themes of the Counter-Enlightenment, and characterized the disillusion with the ideas of the French Revolution. Clausewitz's criticism of the philosophers in Germany who were influenced by Parisian philosophy and politics and 'who have minds which are too distinguished to value local and historical particularities', is again a classic expression of these attitudes. It could have literally been written by Möser, Burke, Adam Müller, or Gentz.[58]

Once more Clausewitz's historical outlook is already revealed in his early writings. As pointed out by Rothfels, it dominated his two works on the Thirty Years War: 'Gustavus Adolphus' and an apparently lost manuscript, 'Views on the History of the Thirty Years War'. Clausewitz consciously chose to deal with a war whose total and devastating nature had terrified the men of the eighteenth century and was regarded by them to be 'inhumane and barbarous'.[59] He interpreted the events in a highly sympathetic manner, revealing the utmost sensitivity to the particular conditions of the period and the concrete challenges that the personalities involved had faced. In contrast to the universal standards employed by the men of the Enlightenment, Clausewitz asserted that the nature of each war depended on the state of the countries and peoples involved, on their customs, political situation, spirit, culture, and so on.[60] Indeed, 'The various great wars constitute many different eras in the history of the art.'[61]

This classic statement of the historicist position was reiterated by Clausewitz on several later occasions. The claim to perfection, he wrote in the essay 'On the State of the Theory of War' is 'one of those boasts with which every period now and again seeks to ornament the events of its day'.[62] Against Bülow's and Jomini's

[58] 'Umtriebe' in Rothfels (ed.), *Schriften*, p. 166. For Clausewitz's political views, see the last section of Ch. 7. II.

[59] 'Gustav Adolph', *Werke*, ix. 19; Rothfels, *Politik und Krieg*, pp. 61–2.

[60] See n. 59 above.

[61] 'Ansichten aus der Geschichte des Dreissigjährigen Krieges'; cited by Rothfels, *Clausewitz, Politik und Krieg*, pp. 61–2.

[62] 'Über den Zustand der Theorie der Kriegskunst', in Schering (ed.), *Geist und Tat*, pp. 52.

universal principles and standards of measurement, he wrote in *On War*:

> It is plain that circumstances exert an influence that cuts across all general principles . . . a critic has no right to rank the various styles and methods that emerge as if they were stages of excellence, subordinating one to the other. They exist side by side, and their use must be judged on its merits in each individual case.[63]

Following in Scharnhorst's footsteps, Clausewitz therefore emphasized the absolute dependence of theory on concrete historical experience. Historical experience is the source of all knowledge, and, in view of the artificial nature of contemporary military theory, it is by far superior to any other study. He concluded his *Principles of War for the Crown Prince* by stressing precisely this point.

Clausewitz's conception of the nature of historical experience and study is most fully presented in *On War*, Book II, chs. 5 and 6. Though this constitutes one of his most interesting analyses, only some of its main points can be cited here. Most historical writing, he maintained, bore witness to an arrogant, dogmatic, and superficial study and judgement of the past. The subjugation of the past to the rule of one-sided systems and principles involved rising above the conditions and individuals peculiar to each particular case, and harnessing a wide, but tendentious, uncircumstantial, and uncritical variety of examples to support abstract conceptions. The purpose of historical study is not to provide doctrines but to train judgement through indirect experience of a profession in which direct experience of sufficient scope is often unattainable. This can only be achieved by intimate familiarity with the conditions of the events studied, even at the expense of concentrating on a selective few historical cases. Furthermore, since the practical purpose of the study of military history is geared to the present, it should focus on modern history. The closer the period is to the present, the more conditions are likely to be similar to it.

It is, therefore, not surprising that Clausewitz's historical works constitute the bulk of his remaining literary works. Seven of the ten volumes of his *Werke* are composed of studies of the great campaigns of modern Europe since the Thirty Years War, particularly the wars of the Revolution and Napoleon. His theoretical work too is

[63] *On War*, VI, 30, p. 516.

characterized not only by many historical analyses and references but also by a strong historical spirit. The most striking example is to be found in *On War*, VIII, 6B. In a few pages Clausewitz offers a most penetrating outline of the transformation of war throughout history as a result of

> the nature of states and societies as they are determined by their times and prevailing conditions . . . The semi-barbarous Tartars, the republics of antiquity, the feudal lords and trading cities of the Middle Ages, eighteenth-century kings and the rulers and peoples of the nineteenth century—all conducted war in their own particular way, using different methods and pursuing different aims.[64]

The perceptive analysis that follows—much richer than implied in the opening passage—will not be cited here. More important is the conception behind it which concludes the narrative:

> Our purpose was not to assign, in passing, a handful of principles of warfare to each period. We wanted to show how every age had its own kind of war, its own limiting conditions and its own peculiar preconceptions. Each period, therefore, would have held to its own theory of war, even if the urge had always and universally existed to work things out on scientific principles.[65]

The last sentences represent the culmination of Clausewitz's historicist conception. Their implications for the possibility of a universal theory of war, as opposed to a theoretical formulation of the conditions peculiar to each time and place, is strikingly Pyrrhonic and destructive. They present, however, only one aspect of Clausewitz's thought. The core of his theoretical work and the major difficulties he encountered in its development were how to formulate a universal theory of war which would be valid despite and within the great diversity of historical experience.

[64] *On War*, VIII, 6B, p. 586. [65] Ibid. 593.

III HOW TO FORM A UNIVERSAL THEORY OF WAR?

Clausewitz's reformulation of the concept of military theory, which was directed against the theoretical outlook of the Enlightenment, was bound up with his effort to devise an adequate military theory of his own. His ideas evolved from general notions during the reform era into a comprehensive and systematic treatise on war written during the period of peace.

In his first works, when he was mainly concerned with developing his attack which aimed at the destruction of the strategic systems, Clausewitz's ideas regarding the possibility of formulating a positive theory of war appear mainly in a negative form. In his critique of Bülow in 1805, Clausewitz was almost unwilling to commit himself on this point. If he were to be asked, he wrote, in the light of the demands that he set for a theory of strategy, whether such strategic theory was at all possible, his reply would be 'that we have neither committed ourselves to write one, nor to prove its possibility, and that we were less inclined to object to the confession: "I do not believe in the art [of war]"' [Berenhorst's], than to the 'Babylonic confusion of language which prevails in military ideas'.[1]

During the reform era a developing shift in Clausewitz's emphases can be traced. Although he was still much concerned with the criticism of contemporary military thinkers, his thoughts were moving forward to more positive problems. In the essay 'On the State of the Theory of War', Clausewitz opposed contemporary military thinkers not because their theory was unhistorical, but because most of them found it difficult to think theoretically, and therefore resorted to examples and eclectic historical discussions. This problem was less severe than might be thought because, according to Clausewitz, history was the basis of theory, and in the absence of adequate theory, historical study was the only possible form of military education. 'However', Clausewitz wrote in the conclusion of the essay, 'none of this keeps us from confessing that we expect great advantage from an intelligent development of theory, partly for the training of young students, and even more for the development of the art itself.' As pointed out by Paret, the programmatic note here is unmistakable.[2]

[1] 'Bülow', in Hahlweg (ed.), *Verstreute kleine Schriften*, p. 82.

[2] 'Über den Zustand der Theorie der Kriegskunst', in Schering (ed.), *Geist und Tat*, pp. 59–60; Paret, *Clausewitz*, p. 156.

Indeed, in the years 1807–9, Clausewitz laid the foundations for his theory of war by developing an intellectual structure that would integrate the diversity of historical experience with a universal approach. In his note 'On Abstract Principles of Strategy' (1808), he briefly surveyed the transformations of the face of war since the Thirty Years War. The result of these transformations was that military theories, which have actually simply reflected changing 'manners' of warfare, have always been invalidated by new historical experience. These changes had been so rapid and far-reaching that 'the books on war have always come out late and in all times they have described dead manners'.[3]

If this is the case, is a universal theory of war possible at all? It is possible, according to Clausewitz, because beyond the diversity of historical experience and the changing 'manners' of each period there exists a universal, constant element, which is the true object of theory. Theory should aim at the 'lasting spirit of war', a concept which already figures prominently in Clausewitz's notes of 1804. The various forms of the art of war decline in time, but the spirit of war escapes change, and must not be 'lost sight of'.[4]

The same conceptual framework is repeated in Clausewitz's letter to Fichte in January 1809, in which Clausewitz criticizes Machiavelli and, implicitly, also Fichte himself for trying to revive the warfare of the ancients. Rather than obsolete 'manners' and 'forms' it is the 'lasting spirit of war' that should be restored.[5]

This conception that suggests an integration of the historical with the universal was deeply rooted in Clausewitz's intellectual environment. Paret pointed out its clear affinity to Schleiermacher's celebrated conception of religion, which attracted much attention during Clausewitz's formative years. Positive religions and ethical systems, wrote Schleiermacher in his famous *Reden über die Religion* (1799) and *Monologen* (1800) appeared in history in a rich variety of forms; they rise and decline but their spirit remains one and universal. Shortly before composing the note on strategy of 1808, Clausewitz wrote to his fiancée:

[3] 'Strategie' (1808), in Hahlweg (ed.), *Verstreute kleine Schriften*, p. 47.

[4] Ibid. 46–7. The similarity to Lossau's ideas is amazing: a systematic theory of war is impossible, and 'thus there can be no lasting textbook for war'. 'War always appears as new; only the spirit of war remains the same.' Lossau, *Der Krieg* p. 35.

[5] Letter to Fichte, Schering (ed.), *Geist und Tat*, esp. p. 77.

Religious feeling in its elemental purity will eternally exist in men's hearts, but no positive religion can last forever. Virtue will eternally exert its beneficial influence on society; but the universality of this global spirit cannot be expressed in the restrictive form of a code of laws, and form itself will shatter sooner or later when the stream of time has washed away or reshaped the surrounding contours.[6]

Schleiermacher's influence here is all too apparent. As we shall see, Clausewitz's conception of the compatibility of the historical and the universal also derived from several other sources and was quite common during the genesis of historicism.

What is the nature of the universal in war? Clausewitz's notes on strategy of 1808–9 reveal the problem that was to figure prominently in his attempts to formulate a theory. The theory of strategy 'allows the setting up of few or no abstract propositions'. One cannot escape the multitude of minor circumstances. 'All the authors that in modern times have sought to treat this part of theory abstractly and philosophically provide a clear indication of this; they are either simply trivial, or they get rid of triviality through one-sidedness.' Venturini belongs to the former category; Bülow and Dumas to the latter.[7] Contemporary military thinkers are criticized, but the theoretical problem preoccupies Clausewitz's own mind. One can either offer clearcut doctrines by ignoring all exceptional conditions, or try to cover all possibilities and provide no positive advice. A priori abstractions always fall between the Scylla of partial validity and the Charybdis of the commonplace.

In the note of 1809, Clausewitz elaborated on this problem:

Formula [is] abstraction. When by the abstraction nothing which belongs to the thing gets lost—as is the case in mathematics—the abstraction fully achieves its purpose. But when it must omit the living matter in order to hold to the dead form, which is of course the easiest to abstract, it would

[6] Letter to Marie, 5 Oct. 1807, Linnebach (ed.), *Briefen*, pp. 142–3, cited by Paret, *Clausewitz*, p. 167. Clausewitz's affinity to Schleiermacher's ideas was briefly pointed out by Weniger, 'Philosophie und Bildung im Denken von Clausewitz', in Hubatsch (ed.), *Schicksalwege*, p. 143, and repeated by Paret. In 1808 Schleiermacher became the professor of theology in the newly founded University of Berlin and one of the major exponents of the awakening Prussian national spirit. As shown by Paret, Clausewitz almost certainly knew him personally at that period. For Schleiermacher's ideas in this connection, see his speeches *On Religion* (London, 1893), esp. speeches 1, 2, 5. Schleiermacher's emphasis on emotion as the constitutive element of religion was, of course, in step with Clausewitz's general affinity with the message of Romanticism.

[7] 'Strategie' (1808), in Hahlweg (ed.), *Verstreute kleine Schriften*, p. 46.

be in the end a dry skeleton of dull truths squeezed into a doctrine. It is really astonishing to find people who waste their time on such efforts, when one bears in mind that precisely that which is the most important in war and strategy, namely the great particularity, peculiarity, and local circumstances, escape these abstractions and scientific systems.[8]

Clausewitz's period of service between 1810 and 1812 as instructor at the military academy and military tutor to the Prussian crown prince undoubtedly stimulated his interest in formulating an adequate positive theory of war. For the first time, he was engaged in teaching, and had to give his views on war a didactic form which culminated in his *Principles of War for the Crown Prince*. However, the beginning of the period of peace was, in this respect, the crucial turning-point in Clausewitz's career. His intellectual activity now became his major preoccupation, focusing on the writing of a comprehensive theoretical work on war. During his period in Koblenz (1816–18), Clausewitz wrote a concise theoretical treatise, the first attempt in a process which was to lead to the writing of *On War*. This early treatise has not survived, but what appears to be its preface and an additional comment that Clausewitz wrote on the treatise's character and composition were included in the posthumous publication of his works. In the preface and comment Clausewitz again put forward the theoretical structure that had emerged in 1807–9.

The 'scientific character' of his work, Clausewitz wrote,

consists in an attempt to investigate the essence of the phenomena of war and to indicate the links between these phenomena and the nature of their component parts . . . the propositions of this book therefore base their inner necessity [*innere Notwendigkeit*] on the secure foundation either of experience or of the concept [*Begriff*] of war as such.[9]

While its surface is in flux, war has an immutable core: its 'spirit', 'essence', 'nature' or 'concept'.

We have already seen the affinity of this conception to that of Schleiermacher. Its formulation in 1816–18 also clearly reveals

[8] 'Strategie' (1809), in Hahlweg (ed.), *Verstreute kleine Schriften*, pp. 60–1.

[9] 'Author's Preface', *On War*, p. 61; whenever I hereafter deviate from the Princeton edition of *On War*, the German original is cited in square brackets. Again compare this with Lossau, another disciple of Scharnhorst whose book had appeared only a few years before: theory aims at 'correct concepts' on the 'nature of war', which 'appear when one develops the concept of war.' Lossau, *Der Krieg*, pp. 2, 6.

Clausewitz's profound debt to Scharnhorst. Theory aimed at correct concepts, had to be grounded in experience or in the nature of war, and pointed out the necessary relations between the parts and the whole; this was the intellectual framework that Scharnhorst had formulated in the 1780s and reiterated throughout his life. As mentioned earlier, Aron was the first to call attention to the striking similarity of this formula to Montesquieu's celebrated definition of laws as 'the necessary relation arising from the nature of things'. Indeed, Clausewitz too, appears to have drawn not solely from Scharnhorst but also from Montesquieu himself. In his comment on the treatise of 1816–18, Clausewitz presents Montesquieu's work as the model that was in his mind when writing his own work.[10] While this reference focuses on structure rather than on content, Clausewitz's conception of the nature of theory and his reference to Montesquieu at the very moment when this conception is elaborated upon and put into practice betray a much deeper affinity. Clausewitz was familiar with, and referred to, Montesquieu's work as a young man, and now, when he turned to write his theoretical treatise, Montesquieu's integration of the historical and empirical on the one hand with the universal on the other appears to have emerged as a model. This affinity should certainly not be exaggerated; to use Clausewitz's words in a wider sense, 'the manner in which Montesquieu dealt with his subject was vaguely in my mind'.[11]

The blending of a high degree of sensitivity to the diversity of historical experience with a belief in certain universal elements is typical of the early period of historicism. Meinecke argued that this blend reflected the legacy of the old tradition of natural law within the historicist outlook which reacted against this tradition.[12] Indeed, the tension inherent in this blend has been pointed out especially in relation to its most classic manifestation in Montesquieu's *Spirit of the Laws*.[13] Returning to Clausewitz, would the notion of a universal

[10] 'Comment', *On War*, p. 63. [11] Ibid.

[12] In this connection see Herder's highly interesting statement: 'The art of war may change with the changes in weapons, times, and state of the world; but the spirit of man—which invents, deceives, conceals its purposes, goes to the attack, defends itself or retreats, discovers the weaknesses of its enemy, and in one way or another uses or misuses them for its advantage—remains at all times the same.' Herder, *Ideen zur Philosophie der Geschichte der Menschheit* (1784–91) XIII, 6; cited by Rothfels, *Clausewitz*, p. 63. Whether Clausewitz was actually aware of this passage is unknown.

[13] See Meinecke, *Historicism*, ch. 3; and Berlin, 'Montesquieu', *Against the Current*, pp. 130–61. For the general inherent tension here see, Reil, *The German Enlightenment and the Rise of Historicism*, ch. VIII, and esp. p. 162.

essence withstand the threat of historical relativity? Would the belief in the 'nature of things' not conflict with the test of 'experience'? In 1816–18 Clausewitz believed that his solution rose to the challenge.

He was still preoccupied, however, with the tendency of universal propositions to lead towards empty formalism, triviality, and truisms. In the preface of 1816–18, he quoted extensively from a work by the famous G. C. Lichtenberg (1742–99), the Göttingen science professor who grew highly sceptical about human knowledge and outlook on the world, and whose aphorisms, published posthumously, were widely read. According to Clausewitz, Lichtenberg's 'Extract from Fire Regulations', satirizing the meticulous, dead formalism of system builders, strikingly fitted the existing military theories.[14] Curiously enough, as we have seen in the note on strategy of 1808, Clausewitz argued the precise opposite against the systems of Dumas, Bülow, and Jomini, his major opponents. They formulated principles which were packed full of content but which were one-sided. In fact, the whole issue reflects a recurring theme in Clausewitz's own mind which also reappears in the comment on the work of 1816–18 and is stated in more personal terms: 'I wanted at all costs to avoid every commonplace, everything obvious.'[15]

Clausewitz approached the writing of *On War* with fairly consolidated ideas on the nature and boundaries of military theory. Like his contemporaries, he believed that the conduct of operations was the true subject of theory, not yet discovered by the early military thinkers of the Enlightenment.[16] However, Scharnhorst's influence and the legacy of Kant's theory of art convinced him that doctrines of absolute applicability for the conduct of war were impossible. The historicist outlook and Schleiermacher's formulation of the traditional message of pietism and Moravianism, which rejected all religious dogmas, positive doctrines, and any other attempt to capture the variety of universal religious feeling in rigid intellectual structures, reinforced this conviction. No rule or principle could cover the diversity of reality nor the different requirements of action. The point made in 1808 is reiterated: 'all the principles, rules, and methods

[14] 'Author's Preface', *On War*, pp. 61–2. For Lichtenberg see J. P. Stern, *Lichtenberg, A Doctrine of Scattered Occasions* (Indiana, 1959).

[15] 'Comment', *On War*, p. 63.

[16] See esp. *On War*, II, 2, pp. 133–4.

will increasingly lack universality and absolute truth the closer they come to being positive doctrine'.[17]

Rules and principles for action are by no means illegitimate in themselves as long as their value and limits are understood correctly. As in Kant's theory of art, their justification is that they provide a way to give the officer some guidance for conduct in war by conceptual means. Hierarchically, they include the very general, such as law—which is too comprehensive and strict a conception to be applicable to anything in war—and progress to principles and rules, directions, regulations, and methods which deal with minute details.[18]

The availability and usefulness of these rules of action diminishes the higher the level of the conduct of war. At the lower levels, in the sphere of tactics, rules of action are easier to formulate because they deal with more physical, material, and technical factors. They are also essential because the enormous number of activities and people involved in these levels require rules, directives, and methods to regulate and unify their operations, render general training possible, and direct rapid and determined action under conditions of shortage of information and time, without the need for rethinking the situation in each individual case. By contrast, at the higher levels of war, in strategy, activity is imbued with subjective factors and conscious decisions, and the issues in point are major and crucial. Here, the commander's free considerations play the decisive part.[19] In any case, all rules for action require circumstantial and critical application involving judgement, and can never be used to criticize opposing decisions and courses of action taken in individual cases and under specific conditions.[20]

In itself, this conception of rules and principles is therefore not very different from that of Clausewitz's predecessors, particularly if one does not accept Clausewitz's caricature of them. However, for Clausewitz, these practical rules and principles could never be considered as the theory of war itself. For that, one had to look elsewhere. The rules and principles merely provided one of the

[17] *On War*, II, 5, pp. 157–8; see again the section, 'A Positive Doctrine is Unattainable', ibid. II, 2, p. 140.

[18] *On War*, II, 4, pp. 151–2.

[19] Ibid. II, 2, pp. 140–1, 147; II, 4, pp. 152–3.

[20] Ibid. II, 4, pp. 151–2, see also II, 5, p. 158; and 'Über Kunst und Kunsttheorie', in Schering (ed.), *Geist und Tat*, pp. 161–2.

bridges—and the crudest one at that—between the need for concrete action and the real theory of war. 'Theory should be study, not doctrine'; it is not a 'manual for action'.[21] The entire military school of the Enlightenment with its rules and principles simply missed the main point: the universal nature of war, its lasting spirit.

What then is the theory of war? The conception that emerged in 1808–9 and 1816–18 reappears. Theory is to be 'used to analyse the constitutive elements of war'.[22] It is 'the field of universal truth that cannot be inferred merely from the individual instances under study'. But it also belongs to the empirical sciences in the sense that 'while, for the most part, it is derived from the nature of things, this very nature is usually revealed to us only by experience'.[23]

Again, if one shrinks from one-sided doctrines, one is in danger of falling into empty formalism. 'When we contemplate all this, we are overcome by the fear that we shall be irresistibly dragged down to a state of dreary pedantry, and grub around in the underworld of ponderous concepts.' Fortunately, in the broad sense, theory is far from being divorced from concrete action. 'Theory cannot equip the mind with formulas for solving problems . . . but it can give the mind insight into the great mass of phenomena and of their relationships, then leave it free to rise into the higher realms of action.'[24] Ultimately, theory is to become capability through critical analysis and practical rules and principles.[25]

This is therefore Clausewitz's conception of theory and his guiding ideal. Above historical study and crude rules there exists a universal theory which reflects the lasting nature of war, transcends the diversity and transformations of past experience, and is both generally valid and instructive. Indeed, it is time to turn to the application of this conception.

In an undated note, written sometime during an advanced stage of the composition of *On War* and describing the state of his work, Clausewitz argued for the feasibility of a universal theory of war, citing a long list of propositions which summarized themes from the

[21] *On War*, II, 2, p. 141.

[22] Ibid.; see also VIII, 1, pp. 577–8.

[23] Ibid. II, 5, p. 157; II, 6, p. 170.

[24] Ibid. VIII, 1, p. 578; compare this with the preface to the work of 1816–18 where Clausewitz wrote that instead of presenting ready-made doctrinal structures (*fertigen Lehrgebäudes*), his work offered material for them; 'Author's Preface', ibid. p. 61.

[25] See esp. *On War*, II, 2, pp. 141 and 147; II, 5, p. 156; VIII, 1, p. 578.

manuscript. Though obviously schematic and ill-organized, this list is highly significant. It is rarely referred to, one dares suggest, because commentators have been somewhat uneasy about its content.

> It is a very difficult task to construct a philosophical [*philosophische*] theory for the art of war, and so many attempts have failed that most people say it is impossible, since it deals with matters that no permanent law can provide for. One would agree, and abandon the attempts, were it not for the obvious fact that a whole range of propositions can be demonstrated without difficulty: that defence is the stronger form of fighting with the negative purpose, attack the weaker form with the positive purpose; that major successes help bring about minor ones, so that strategic results can be traced back to certain turning points; that a demonstration is a weaker use of force than real attack, and that it must therefore be clearly justified; that victory consists not only in the occupation of the battlefield, but in the destruction of the enemy's physical and psychological forces, which is usually not attainable until the enemy is pursued after a victorious battle; that success is always greater at the point where the victory was gained, and that consequently changing from one line of operations, one direction, to another can at best be regarded as a necessary evil; that a turning movement can only be justified by general superiority or by having better lines of communication or retreat than the enemy's; that flank positions are governed by the same consideration; that every attack loses impetus as it progresses.[26]

How universal are these propositions and how successful are they in escaping the dilemma of one-sidedness and triviality? This is perhaps better left to the reader's own consideration. Rather than discussing these propositions eclectically, the following chapter will attempt to trace the development of Clausewitz's central line of thought on the nature of war.

[26] 'Undated Note', *On War*, p. 71.

7

Clausewitz

The Nature of War

I MILITARY DECISIVENESS AND POLITICAL GREATNESS: THE NAPOLEONIC MODEL

The nature of war is fighting; hence all the characteristics of its 'lasting spirit': the primacy of the engagement and of the major battle, aided by a massive concentration of forces and aggressive conduct, and aiming at the total overthrow of the enemy. Throughout his life, this conception was the centre-piece of Clausewitz's military outlook. It reflected the overwhelming impact of the Napoleonic experience, was the source of Clausewitz's attacks on the war of manœuvre in all periods and particularly in the eighteenth century, and formed the basis for his belief in a universal theory of war.

Ironically, in 1827, this whole military outlook fell into a deep crisis. In the middle of composing *On War*, Clausewitz's line of thought underwent a drastic change of direction, the only revolutionary transformation in the otherwise steady evolution of his ideas. In a note on the state of his work dated 10 July of that year, Clausewitz announced his intention to revise *On War* on the basis of two guiding ideas: firstly, that there are two types of war: all-out war and limited war; and secondly that war is the continuation of policy by other means.

The crisis of his conception of the nature of war was equally destructive for Clausewitz's lifelong conception of theory. In his efforts to resolve this comprehensive crisis, he transformed but *did not* abandon his old military outlook, and resorted to completely new theoretical devices. He was preoccupied with this during his last three working years.

Unfortunately, the origins and nature of Clausewitz's new theoretical framework have remained a mystery, and consequently, the exact nature of the transformation in his thought has not been entirely clear either. This explains why Clausewitz's ideas could be interpreted so differently by successive generations. Whereas the men of the nineteenth century emphasized the place of the major battle and the element of destruction in Clausewitz's thought, modern readers, contending with the problem of limited war and seeking out the full complexity of the link between political and military activity, have stressed themes in his later thought. As this has been coupled by a strong reaction, particularly in Germany, against the military and political legacy of the German Reich, a new, 'good' Clausewitz has had to be created, set apart from his 'bad' successors. While blaming their discredited predecessors for being tendentious and one-sided, modern interpreters have therefore themselves failed to recognize that the imperative of destruction was the basis of Clausewitz's conception of war. As we shall see, some have even denied that he held such an idea at all. The obscurity of Clausewitz's text has continually left room for conflicting and unhistorical interpretations.

Clausewitz's conception of the nature of war stemmed from both his military and political outlook, and was incorporated into his definition of war. In the military sphere this conception reflected the earth-shattering collapse of the warfare of the *ancien régime* when confronted by the Revolutionary and Napoleonic art of war. With the emperor's great triumphs of 1805–7 this process was completed. For the first time in the history of modern Europe a single state had inflicted a crushing defeat over all the other powers of the continent. Eighteenth-century warfare, which, because of the political and social structure of the *ancien régime*, had been relatively limited in aims and scope was now increasingly discredited and perceived as inadequate, if not absurd.

This upheaval was not, of course, expressed solely by Clausewitz but underlay almost the whole of military thought at the turn of the nineteenth century. We have already seen it reflected in varying degrees in the transition from Bülow and Archduke Charles to Jomini. The total mobilization of forces, initiative, aggressiveness, and rapid decision in battle now dominated warfare. Yet, nowhere was the reaction against the past and the embracing of the new spirit of war as powerful as in the defeated Prussia. And of all of Clausewitz's contemporaries no one gave the new trends more far-reaching expression—a fact which, until our own times, was obvious to everyone.

Closely linked to Clausewitz's military outlook were his political attitudes. Again interpreters have not paid them sufficient attention and failed to appreciate their interrelation with Clausewitz's theoretical work, ignoring his actual historical background and intellectual career. Clausewitz saw the face and map of Europe radically altered by determined and powerful political and military activities and witnessed his country, which dabbled in diplomatic manœuvres and military half-measures, lose its independence and status as a great power in a single powerful blow. To these experiences were added the dynamic and vitalistic effect of Romanticism and the fervent energy and feeling generated by rising nationalism. Clausewitz urged the state to pursue great objectives, to be determined in its actions, and to put the utmost power behind them. It is not surprising that these notions, as well as Clausewitz's military views found support and reinforcement in Machiavelli's works.

Paret points out the major themes behind Clausewitz's enthusiastic interest in and warm appreciation of Machiavelli, which is well documented in Clausewitz's early works:[1] the emphasis on the moral energies that animate nations and armies, and the comprehensive and penetrating presentation of politics, in the centre of which stands the role and skilful use of force.[2] To understand the full scope of Clausewitz's attraction to Machiavelli it is necessary, however, to note the fascinating parallels in their historical and psychological position, both in the military and the political spheres. There is a surprising similarity in the developments they witnessed and in their reactions to them.

In the military sphere Machiavelli saw the weakness of the mercenary armies, the *condottieri*, with their cautious tactics, fully exposed by the emergence of the new vigorous national armies of Switzerland, France and Spain, and he called for the creation of a civic militia, motivated by national sentiments. Clausewitz witnessed the collapse of the professional armies of the *ancien régime* when

[1] 'Strategie' (1804) in Hahlweg (ed.), *Verstreute kleine Schriften*, p. 9, presenting Machiavelli as having 'a very sound judgement in military affairs'; Rothfels (ed.) *Schriften*, p. 63: 'no book in the world is as essential for the politician as Machiavelli's'. For other references see 'Strategie' (1804), articles 4, 5, 6; 'Historisch-politische Aufzeichnungen' (1805) and 'Bei Gelegenheit der russichen Manifeste nach dem Tilsiter Frieden', in Rothfels (ed.), *Schriften*, pp. 4 and 62 respectively; and Clausewitz's letter to Fichte, in Schering (ed.), *Geist und Tat*, p. 77.

[2] Paret, *Clausewitz*, pp. 169–79.

confronted by the aggressive armies of mass conscription raised by Revolutionary France, and he too demanded the creation of an army of general conscription supported by a national militia. Throughout his life, Clausewitz, following in Machiavelli's footsteps, denounced the era of the *condottieri*, as well as the warfare of the *ancien régime*, as a degeneration of the art of war.[3]

In the political sphere, Machiavelli witnessed the eclipse of the once-proud Italian city-states and the impotence of their diplomacy in contrast to the real political and military might of the new powers. He stressed the dominance of force in politics and called for a dynamic political and patriotic revival. Clausewitz, as mentioned earlier, saw the diplomatic manœuvres of the mediocre heirs of Frederick the Great stripped of all their efficacy by Napoleonic power. After the disaster, he stood out, even in the reform circle, in his call for a bold and determined policy, and in his relentless search for every opportunity—the Spanish guerrilla warfare, the Austrian war of 1809, the French invasion of Russia—to launch a total war of independence, even if it might lead to destruction. His bitter and fierce criticism of his country during this period is clearly marked by Machiavellian themes: contempt for half-measures, indecisiveness, and inactivity which, in the end, are bound to lose all worlds. In his defence of Machiavelli, Clausewitz wrote: 'Chapter 21 in Machiavelli's "Prince" [warning against neutrality and calling for rallying with one of the sides] is the code for all diplomacy, and woe to those who distance themselves from it!'[4] Activity, vitality, and power in the political as well as in the military spheres were the essence of Clausewitz's outlook.

This outlook was incorporated into Clausewitz's conception of the nature of war: 'Essentially war is fighting, for fighting is the only effective principle in the manyfold activities generally designated as war.' The developments in weapons 'brought about great changes in the forms of fighting. Still no matter how it is constituted, the concept of fighting remains unchanged.'[5]

[3] See e.g. *On War*, II, 6, p. 174; VIII, 3, p. 587.

[4] Rothfels (ed.), *Schriften*, p. 64. For Clausewitz's national fervour, plans of insurrection against the French, and criticism of his country's policies, see esp. Paret, *Clausewitz*, chs. 8.I, 8.IV.

[5] *On War*, II, 1, p. 127. Compare Schleiermacher's speeches *On Religion*, for example: 'everything called by this name [religion] has a common content', religious feeling (p. 13); 'The essential oneness of religiousness spreads itself out in a great variety' (p. 50).

We have finally reached the actual content of Clausewitz's theoretical conception, which was unveiled in his letter on religion, note on strategy, and letter to Fichte. Whereas the 'forms' of war are diverse and changing, its 'spirit' is universal. Like religious feeling in religion, fighting is the constitutive element of war, which allows us to regard the many different wars as part of a single phenomenon. This, however, is not merely a statement defining the common denominator of all wars; military theory which is blind to, or evades, the imperative of fighting—as the thinkers of the eighteenth century allegedly did—creates a false picture of war, which is bound to lead to disaster. The dominance of fighting determines the whole character of war. From his earliest works Clausewitz stressed this point.

In the notes on strategy of 1804, the full scope of Clausewitz's military outlook, inextricably linked to his political state of mind, is unfolded for the first time. Firstly, Clausewitz rejects the limited warfare of the eighteenth century and denounces the central role of fortresses, the division of armies, and Fabian strategy.[6] The correct conduct of war is diametrically opposed 'I would not like to print this, but I cannot hide from myself that a general cannot be too bold in his plans, provided that he is in full command of his senses, and only sets himself aims that he himself is convinced he can achieve.'[7] In a nutshell, 'the art of war tells us: *go for the greatest, most decisive purpose you can achieve; choose the shorest way to it that you dare to go*'.[8] '*War should be conducted with the utmost necessary or possible degree of effort.*'[9] One should achieve the utmost concentration of force, and strike the enemy with the maximum power. Defence ought to be adopted only if one is too weak to attack. The enveloping strategic manœuvre against the enemy's rear, recommended by Bülow with the approval of Venturini, Dumas, and Massenbach, has indeed the advantage of threatening the enemy's communications; but its success is dubious, and direct action from a central position is more effective. Frederick the Great's conduct in the Seven Years War, Napoleon's Italian campaign of 1796, and many other examples from the history of war attest to this. Clausewitz worked out a strikingly similar conception to the one that Jomini developed that very same year but had not yet published.[10]

[6] 'Strategie' (1804), s. 9, in Hahlweg (ed.), *Verstreute kleine Schriften*, pp. 14–16.
[7] Ibid., s. 12, pp. 19–20.
[8] Ibid., s. 12, p. 19.
[9] Ibid., s. 13, p. 21.
[10] Ibid., s. 22, pp. 35–6; s. 13, pp. 24–5; s. 15, pp. 27–9; s. 19, p. 33.

All this derived from the main point: 'It can be absolutely universally said: all that demands the use of military forces has the idea of the engagement at its base.' Engagement is the centre of war, toward which all efforts are directed. The belief of Bülow and his fellow-thinkers that victory could be won by means of a brilliant strategic manœuvre, without resorting to battle, was an absurd illusion. Manœuvres are, of course, necessary, but only to achieve favourable conditions for the engagement. Even when the engagement itself does not take place, its expected outcome regulates the conduct of the belligerents like the effect of 'cash on credit in commerce'. 'Strategy works with no materials other than the engagement.' Thus, whereas tactics is defined as the 'use of military forces in the engagement', strategy is but the 'use of individual engagements to achieve the aim of the war'.[11]

This military outlook went hand in hand with *corresponding* attitudes to the political aims of war, and the relationship between policy and war. War is fought for the attainment of a political purpose, 'the purpose of war'. And in 1804 this purpose was also formulated in radical and aggressive terms: either to destroy the enemy's state or to dictate the terms of peace.[12] Among Clausewitz's interpreters who looked upon these alternatives through the prism of the intellectual revolution of 1827, Aron was the only one to note that in 1804 the choice was *not* between total and limited war. Dictated peace terms implied bringing the enemy to his knees. Indeed, *both* options are cited explicitly in 1827 under the single aim of completely overthrowing the enemy.[13] This crushing political purpose is matched by the objective of the military operations, 'the purpose in war', which is 'to paralyse the enemy forces'. 'The destruction of the enemy's armed forces is the immediate purpose of war, and the most direct way to it always constitutes the rule for the art. This destruction can be achieved by occupying his country or annihilating his war provisions or his army.'[14] In considering the 'purpose of operations' one should 'always choose the most difficult, *for this is the one most closely related to the spirit of the art of war*'.[15]

[11] 'Strategie' (1804), s. 20, p. 33; s. 21, p. 35.

[12] Ibid., s. 13, p. 20.

[13] Aron, *Clausewitz*, pp. 88–9. The term 'two types' (*doppelte Art*) which is repeated in 1827, added to this confusion.

[14] 'Strategie' (1804), s. 13, pp. 20–1.

[15] Ibid., s. 12, p. 20.

It is therefore misleading to assume that in 1804 Clausewitz had already been aware of the range of political aims and objectives, and that in 1827 he simply elaborated on it or became fully aware of its implications for the conduct of war. There was a perfect harmony in 1804 between Clausewitz's political and military convictions; *both* were formulated in radical terms. Total concentration of force, the imperative of fighting for decision—these were Clausewitz's conceptions of the nature of war within the context of his general political outlook which called for determined action and great objectives.

This military outlook and political state of mind are again fully revealed in Clausewitz's next comprehensive work, *Principles of War for the Crown Prince* (1812). Clausewitz sent the work to the prince when he left Prussia to join the Russian army, and in it he made a special effort to impress upon the young prince in that critical hour the fervour of his outlook on politics and war. In the programmatic passage that concluded the work he wrote to the prince: 'A powerful emotion must stimulate the great ability of a military leader . . . Open your heart to such emotion. Be audacious and cunning in your plans, firm and persevering in their execution, determined to find a glorious end.'[16]

The work itself reiterates all the themes raised in 1804:

> We always have the choice between the most audacious and the most careful solution. Some people think that the theory of war always advises the latter. That assumption is false. If the theory does advise anything, it is the *nature of war* to advise the most decisive, that is, the most audacious.[17]

The aims in the conduct of war are (*a*) to conquer and destroy the armed forces of the enemy; (*b*) to take possession of the resources of his army; and (*c*) to win public opinion.[18] These aims, it must be noted, are *not* alternative but complementary; they are intended to secure the complete defeat of the enemy.

The first principle of the art of war is the concentration of force, supported by dynamic and determined conduct which avoids half-measures.[19] A defensive position should only be adopted as a means

[16] *Principles*, p. 69.
[17] Ibid. 13–14; my emphasis.
[18] Ibid. 45.
[19] Ibid. 12, 17–19, 21, 46.

for attacking the enemy from a position of advantage.[20] The engagement is the focal point of war, much more important than the skilful combination of engagements (that is strategy).[21] Thus, in war, direct, crushing operations from a central position are superior to concentric enveloping manœuvres; Jomini was right about this while Bülow indulged in illusions. Clausewitz even goes so far as to make the fantastic statement (after Marengo, Ulm, and Jena, to name only the most important examples) that 'Napoleon never engaged in strategic envelopment'.[22]

All this is enough to show that Clausewitz's conceptions were clearly a particular reflection of Napoleonic warfare as perceived in its peak years. This was precisely how Berenhorst saw them when he read Clausewitz's *Principles*:

> The most significant parts of his wisdom, he abstracted from the wisdom, the actions, and the maxims of Napoleon. Indeed, his relationship to Carnot's and Napoleon's method or system of war, today's art of war, is like the relationship of Reinhold, Kiesewetter, and Berg to the philosophy of Hume and Kant [that is, they interpret and populize it] . . . He certainly has the merit of explaining the new art of war very well and intelligently, and he should be recognized as the first to have done so.[23]

Characteristically, Berenhorst regarded Clausewitz's ideas simply as a penetrating interpretation of the particular form of warfare that then prevailed.

Finally, the same themes and view of war are expressed in the early and unrevised parts of *On War*, Books II–VII. They will be briefly noted again, if only to establish the clear continuity in Clausewitz's outlook:

> The very concept of war will permit us to make the following unequivocal statements: 1) Destruction of the enemy forces is the overriding principle of war . . . 2) Such destruction of forces can *usually* be accomplished only by fighting. 3) Only major engagements involving all forces lead to major success. 4) The greatest successes are obtained where all engagements coalesce into one great battle.[24]

[20] *Principles*, p. 17.

[21] Ibid. 15.

[22] Ibid. 48–9.

[23] Berenhorst to Valentini, 1 Nov. 1812, Bulow (ed.), *Nachlasse*, ii. 353. Rather than revealing an affinity with Clausewitz's way of thinking (as suggested by Paret, *Clausewitz*, p. 205), these words stood, in fact, in stark contrast to it; Clausewitz did not regard his ideas as a mere expression of a particular form of warfare.

[24] *On War*, IV, 11, p. 258.

Destruction should be the aim in each individual engagement.[25] Defence is indeed stronger than attack and thus it is the weapon of the weak; but once the defender gains the advantage, he must revert to the attack. The second part of this formula and Clausewitz's full meaning have been missed by most modern commentators who characteristically contrasted his wisdom with his successors' mania for the attack.[26] Again, to stress the link between Clausewitz's military and political outlook: the 'very destruction of the enemy's forces is also part of the final purpose [of war]. That purpose itself is only a slight modification of that destructive aim.' Ignoring this point was at the root of all the false military theories before the Napoleonic Wars.[27]

Even when the engagement does not take place, the very threat of it regulates the conduct of the belligerents. Any other military objective—the occupation of provinces, cities, fortresses, roads, and bridges, the seizure of ammunition, and so on—must merely be seen as an intermediate means intended to achieve a greater advantage for the engagement.[28] The same applies to strategic manœuvres. First, it is interesting to see that as the campaigns of 1813 and 1814 cast doubt on the conception of interior lines, Clausewitz withdrew from his own unequivocal position of 1804 and 1812. In principle, he writes, no a priori advantage can be attributed to either interior or exterior lines; the choice between them is dependent upon the type of warfare and upon circumstances.[29] In either case, he maintains as before, that the strategic manœuvre is secondary in importance to the engagement, and must be subservient to it:

Admittedly, an engagement at one point may be worth more than at another. Admittedly, there is a skillful ordering of priority of engagements in strategy . . . We do claim, however, that direct annihilation of the enemy's forces must always be the *dominant consideration*. We simply want to establish this dominance of the destructive principle . . . one should not swing wider than latitude allows . . . rather than try to outbid the enemy

[25] Ibid. IV, 3, p. 229.

[26] For a fuller analysis of the real nature of this highly interesting relationship, the subject of Bks. VI and VII, see my article 'Clausewitz on Defence and Attack', *Journal of Strategic Studies*, X (March, 1988).

[27] *On War*, IV, 3, esp. p. 228.

[28] Ibid. III, 1, pp. 181–2.

[29] Ibid. VII, 13, pp. 541–2.

with complicated schemes, one should, on the contrary, try to outdo him in simplicity.[30]

Ulm was an exceptional event; the decisive battle is the dominant feature of war.[31]

This outlook encompasses not only the manœuvre, but also every other military means other than the engagement itself. Once Clausewitz's starting point is understood, all his military ideas become crystal clear. For all the significance of surprise, Clausewitz maintains that 'by its very nature [it] can rarely be *outstandingly* successful. It would be a mistake, therefore, to regard surprise as a key element of success in war.'[32] The same applies to cunning. For all its importance and 'however much one longs to see opposing generals vie with one another in craft, cleverness, and cunning, the fact remains that these qualities do not figure prominently in the history of war . . . The reason for this is obvious . . . strategy is exclusively concerned with engagements.'[33] The truly important factors are superiority of numbers and concentration of force at the decisive point.[34]

Hence also Clausewitz's exclusion of all preparatory activities (as well as the supporting services such as maintenance, administration, and supply) from the theory of war proper, which has occasionally surprised commentators. Theory only takes these activities into account as influencing conditions, because, strictly, it 'deals with the engagement, with fighting itself'. Marches, camps, and billets only narrowly escape the same fate, because 'in one respect [they] are still part of combat'.[35]

All these notions lead to Clausewitz's surprisingly dull description of contemporary battle. No manœuvres or stratagems are portrayed. The only image conveyed is of a direct, grey clash of physical and moral masses.[36]

The men of the nineteenth century, the heyday of the idea of all-out war, elevated Clausewitz to the pantheon of classics for his outlook on war described above. But for the present-day reader, after the collapse of the dogma of destruction in the First World War and

[30] *On War*, IV, 3, pp. 228–9.
[31] Ibid. IV, 11, p. 260.
[32] Ibid. III, 9, p. 198.
[33] Ibid. III, 10, p. 202.
[34] Ibid. III, 11, p. 204.
[35] Ibid. II, 1; for the citations, see pp. 132, 129.
[36] Ibid. IV, 2, p. 226.

the renaissance of limited war in the nuclear age, this outlook should have raised questions had its real nature not been obscured by Clausewitz's later development and the difficulties of interpreting it.

However, already at the beginning of the twentieth century when the conception of all-out war still reigned, Camon, one of the most important interpreters of the Napoleonic art of war, argued that Clausewitz misunderstood the essence of Napoleonic strategy, particularly the key manœuvre against the enemy's rear, *la manœuvre sur les derrières*.[37] Jomini's analysis of the Napoleonic art of operations, as opposed to the full context of Napoleonic warfare, was perhaps more concrete and realistic. Indeed, Clausewitz's conception of the Napoleonic art of war was largely a myth, born out of Prussia's traumatic experience and reflecting the prevailing emphasis on moral energies. To draw an otherwise most unlikely parallel, Clausewitz was in a way the theoretical counterpart of the action-hungry Field Marshal Blücher.

Hans Delbrück, the well-known military historian, raised the theoretical problem itself in the late nineteenth and early twentieth centuries when he questioned the universal validity of all-out war, paradoxically relying on Clausewitz's later conceptions. He advocated the legitimacy of the strategy of attrition, and started a celebrated but hardly successful debate in which he was attacked by Theodor and Friedrich von Bernhardi and by Colmar von der Goltz who represented the established strategic convictions of the time. Limited strategy, Delbrück maintained, such as that of the eighteenth century, had to be understood as the natural outgrowth of the particular conditions of the period.[38]

The deep crisis of the idea of all-out war and the direct attack on Clausewitz did not, however, take place until after the traumatic experience of the First World War. Liddell Hart, the most renowned and influential leader of the reaction against the military tradition of the nineteenth century, rehabilitated the discredited warfare of the eighteenth century and very sharply criticized (albeit somewhat superficially and tendentiously) Clausewitz and his legacy.[39] In

[37] H. Camon, *Clausewitz* (Paris, 1911); this work provides a close scrutiny of Clausewitz's histories of Napoleon's campaigns.

[38] For a summary of the debate see Delbrück, *History*, iv. 378–82.

[39] See esp. B. H. Liddell Hart, *The Ghost of Napoleon* (New Haven, 1934).

opposition to Bülow, Clausewitz had said that the aim of strategy was merely to achieve the most favourable conditions of time and place for the battle, and to make use of its results. One could argue, he wrote, that the perfection of strategy was therefore to achieve such favourable conditions as to render battle unnecessary. But, in fact, in reality it was usually advisable to count on fighting. If a general could not rely on the determination of his troops to fight, he would find himself continuously inferior.[40] Completely unaware of this argument, Liddell Hart again reversed the outlook on war; 'even if a decisive battle be the goal,' he wrote, 'the aim of strategy must be to bring about this battle under the most advantageous circumstances . . . The perfection of strategy would be, therefore, to produce a decision without any serious fighting.'[41] Clausewitz and Liddell Hart each interpreted the same logic in terms of the warfare of their times and arrived at diametrically opposed conclusions. Military reality and theory completed a full circle between Bülow and Liddell Hart.[42]

The main problem raised by Clausewitz's military outlook, and most of the reactions described here, have been well presented by Aron: 'Did Clausewitz's antidogmatism degenerate into a new dogmatism?'[43]

Having seen in the previous chapters his conception of theory and criticism of his predecessors, the full irony of Clausewitz's conception of war should be clear. He, who passionately believed that his predecessors' theoretical approach and view of war were totally false, was convinced that he had the key to the true nature of war. The paradoxical result of this conviction was that some of his principal arguments against his predecessors boomeranged. His outlook on war was, in its own way, no less one-sided, dogmatic, prescriptive, and unhistorical.

This puzzling discrepancy can only be understood against the background of the dominant role that Napoleon's crushing warfare played in the period's consciousness and in the discrediting of the

[40] 'Bülow', in *Hahlweg* (ed.), *Verstreute kleine Schriften*, pp. 78–9.

[41] B. H. Liddell Hart, *Strategy, the Indirect Approach* (London, 1954), 338.

[42] That the validity of Clausewitz's criticism of Bülow's attitude to manœuvre and battle is not as obvious as maintained by the German military school was also cautiously pointed out by E. A. Nohn, 'Der unzeitgemässe Clausewitz', *Wehrwissenschaftliche Rundschau*, V (1956).

[43] Aron, *Clausewitz*, p. 85.

old conduct of war, particularly in Prussia.[44] Clausewitz who was one of the major exponents of the new trends gave them logical expression with his definition of war as fighting, interpreted in an expansive and imperative manner. In Clausewitz's eyes this was not one-sidedness and dogmatism but at last the true universal nature of war and consequently the key to its proper conduct.

From this conviction derives also the prescriptive aspect of theory. Much has been written to the effect that Clausewitz totally rejected prescriptive theory, and as we have seen, this interpretation does have roots in Clausewitz's conception of theory. However, this is only a partial understanding of his approach as a whole. He maintained that the theory of war was not prescriptive only in the sense that any doctrine derived from it would always be partial and require judgment in application. But he did believe that the true theory of war provided lessons which the general had to bear in mind. Theory was by no means divorced from praxis; on the contrary, it had to be translated into praxis. Now we have also seen what concrete ideas he had in mind: to aim for great objectives, to achieve the utmost concentration of force, to act as aggressively as possible in order to annihilate the enemy army in a major decisive battle, and to destroy the ability of the enemy state to resist. He believed that 'unnecessary' manœuvres, preference for indirect military means, and evading decision in battle contradicted the spirit of war, were bound to lead to failure, and thus had to be avoided. These ideas are highly imperative; Clausewitz had no interest in empty truths.[45]

These strong convictions regarding the fundamental and universal nature of war also overshadow Clausewitz's historical sensitivity.

[44] See the remarkable similarity of Clausewitz's ideas to those of yet another pupil of Scharnhorst and a fellow-student of Clausewitz in the Institute, Rühle von Lilienstern, a man of vast intellectual interests, an intimate of Adam Müller and Heinrich von Kleist, and a friend of Goethe, Gentz, and the Schlegels. Rühle opened his *Handbuch für den Offizier*, a rev. ed. of Scharnhorst's work (Berlin, 1817), with the following statements: 'The engagement is the principal element of war . . . war is fighting' (p. 1); war is battles chained together or one great battle with intermissions (pp. 1, 435). Rühle, whose military career paralleled that of Clausewitz, also stressed the dominance of moral forces and the need for a theory of war which is rooted in reality and the nature of war (esp. pp. 438–44). For Rühle, see Hagemann, *Von Berenhorst zu Clausewitz*, pp. 55–66; Louis Sauzin's introd. to *Rühle von Lilienstern et son apologie de la guerre* (Paris, 1937); and Paret, *Clausewitz*, pp. 272, 314.

[45] Here too compare with Lossau's *Der Krieg*, p. 2: 'From these concepts [of the nature of war] there must emerge clearly what war is, what the warrior must want, and how one should study war in time of peace.'

In opposition to Jomini in particular, Clausewitz stressed the diversity of historical experience, and asserted that the theoretician must not elevate himself above the times by the force of standards of measurement which he regarded to be universal. Every period's particular form of warfare stemmed from its unique political, social, cultural, and personal conditions. As we have seen, he concluded his historical description of the transformations of the art of war in the context of the particular conditions of each period as follows: 'Each period, therefore, would have held to its own theory of war.' Indeed, this was the pinnacle of Clausewitz's historicism yet also its boundary; the next sentences limit the relativism implicit in this historical view:

> But the conduct of war [*Kriegführung*], though conditioned by the particular characteristics of states and their armed forces, must contain some more general—indeed, a universal—element with which every theorist ought above all to be concerned. The age in which this postulate, this universally valid element was at its strongest was the most recent one.[46]

The wars of the Revolution and Napoleon revealed the nature of war as fighting and a clash of forces, and dispelled the false conceptions which prevailed in various periods in the past.[47] Here too, as throughout his life, Clausewitz treats the warfare of the *condottieri* and that of the *ancien régime* not as genuine expressions of the conditions peculiar to their times, but as a grotesque distortion of the nature of war, necessarily leading to collapse. For all his criticism of Jomini, Clausewitz himself turned the warfare of his own period into a universal yardstick and employed it to dismiss the warfare of complete historical periods, disregarding their internal, circumstantial logic.

All this can only be understood in the spirit of Jomini's bitter complaint that Clausewitz's 'first volume [Books I–IV of *On War*; Jomini clearly referred mainly to Book II] is but a declamation against all theory of war, whilst the two succeeding volumes [the rest of *On War*], full of theoretic maxims, proves that the author believes in the efficacy of his own doctrines, if he does not believe in those of others'.[48] Clausewitz was convinced that in contrast to his predecessors' arbitrary and misleading systems, he himself had

[46] *On War*, VIII, 3B, p. 593.
[47] Ibid.
[48] Jomini, *Summary*, p. 15.

discerned the true nature of war, manifested in the genius of Napoleon, 'the God of War'.[49]

It was nevertheless Clausewitz's sensitivity to the diversity of historical experience that, in 1827, when most of *On War* was already drafted, finally led to the crisis in his outlook on war and conception of theory. The first realization of a problem emerges toward the end of Book VI, 'Defence'.[50] This is no coincidence. Since the aim of defence is to preserve the status quo, the defender may choose to delay operations, withdraw, and avoid confrontation in the hope of wearing the enemy down. This may lead to what Clausewitz called a 'war of observation': a prolonged, indecisive struggle which lacks energy and involves almost no fighting. In truth, the attacker too, sometimes appears to 'ignore the strict logical necessity of pressing on to the goal'.[51] This realization leads to a wider one:

> There is no denying that a great majority of wars and campaigns are more a state of observation than a struggle for life and death—a struggle, that is, in which at least one of the parties is determined to gain a decision. A theory based on this idea could be applied only to the wars of the nineteenth century.[52]

Indeed, 'To be of any practical use', theory must take into account that, apart from 'the kind of war that is completely governed and saturated by the urge for a decision—of true war', there exists a second kind of war.[53] Moreover, 'the history of war, in every age and country, shows not only that most campaigns are of this type, but that the majority is so overwhelming as to make all other campaigns seem more like exceptions to the rule'.[54]

Clausewitz's view of the nature of war as all-out fighting, centring on the engagement, fell into crisis. The note that Clausewitz wrote on 10 July 1827 heralded the celebrated transformation in his thought with which he was to struggle in the writing of Book VIII and revision of Book I of *On War*, and which will be described in the next section.

[49] *On War*, VIII, 3B, p. 593.
[50] Michael Howard, *Clausewitz* (Oxford, 1983), 47, 58.
[51] *On War*, VI, 30, p. 501.
[52] Ibid. VI, 28, pp. 488.
[53] Ibid. VI, 28, pp. 488–9.
[54] Ibid. VI, 30, p. 501.

The devastating effect of this crisis on Clausewitz's conception of theory must first, however, be elucidated:

> One might wonder whether there is any truth at all in our concept of the absolute character of war were it not for the fact that with our own eyes we have seen warfare achieve this state of absolute perfection. After the short prelude of the French Revolution, Bonaparte brought it swiftly and ruthlessly to that point . . . Surely it is both natural and inescapable that this phenomenon should cause us to turn again to the original [*ursprünglichen*] concept of war with all its rigorous implications. Are we then to take this as the standard, and judge all wars by it, however much they may diverge? Should we deduce all the *demands* [*Forderungen*] of theory from it? . . . [Then] our theory will everywhere approximate to logical necessity, and will tend to be clear and unambiguous. But in that case, what are we to say about all the wars that have been fought since the days of Alexander—excepting certain Roman campaigns—down to Bonaparte? . . . We would be bound to say . . . that our theory, though strictly logical, would not apply to reality.[55]

This dilemma shatters Clausewitz's lifelong conception of theory. 'Is one war of the same nature as another?', he asked in a note in which he wrote down the new problems in an attempt to clarify his thoughts.[56] 'All *imperatives* inherent in the concept of war seem to dissolve, and its foundations are threatened.'[57]

Having seen in the previous chapter the development of Clausewitz's conception of theory, the crisis into which this conception fell ought now to be clear: theory conflicted with reality; the 'concept of war' did not withstand the 'test of experience'; the universal contradicted the historical; the unity of the phenomenon of war, based on a 'lasting spirit' that encompassed the diversity of 'forms', disintegrated; and the practical imperatives derived from this 'spirit'—the significant content of theory—lost their validity.

[55] *On War*, VIII, 2, p. 580; the emphasis on the prescriptive aspect of theory is mine.

[56] Schering (ed.), *Geist und Tat*, p. 309; Aron, *Clausewitz*, p. 101.

[57] *On War*, VIII, 6A, p. 604; the emphasis on the prescriptive aspect of theory is mine.

II POLITICS AND WAR: THE AMBIGUOUS TRANSFORMATION CLARIFIED

The relationship between politics and war dominated Clausewitz's thought during his last years, generated a revision in his theory of war, and has attracted most of the attention in our time. This subject was presented by Clausewitz—for reasons which will be dealt with—as a single whole, all the elements of which were closely bound together. However, three major ideas can be discerned here, whose origins, development, and content, though not unrelated, were separate and distinct: (*a*) war as an extension of its social milieu, an idea that reflected the historicist message; (*b*) the diverging scope of political aims and military operations; (*c*) the state as the highest and unifying expression of human life and the guardian of political and moral ends, logically governing the military body; this idea reflected the Prussian traditional *raison d'état* and the formative stage of what was to be known as the German conception of the state.

War and the Social Milieu: Applying the Historicist Message

War is an integral part of comprehensive social and political reality which shapes its particular characteristics in any given period; out of all of Clausewitz's ideas on politics and war this was the one that played a dominant role in his thought from his youth. He absorbed it from the rising conceptions of historicism and directly from Scharnhorst. In fact, in the modern sense, this idea is concerned with the relationship of war to society rather than to politics. Most of the following themes have already been discussed throughout this work, particularly in relation to Scharnhorst's and Clausewitz's historical outlook, and will therefore be only briefly reviewed here.

As we have seen, Montesquieu's *Spirit of the Laws* revealed to the men of the eighteenth century a new depth of affinity between the array of elements and circumstances which made up any given historical fabric. Geographical and economical conditions, social structure, legal and political systems, religious faith and institutions, and cultural forms were intertwined in a diversity of particular manifestations. This highly influential idea left its mark on the military thinkers of the Enlightenment, and was extensively applied to the field of war by Guibert and Lloyd as well as by Jomini in his later works. However, its impact was restricted by the pronounced universalism of the military thinkers of the Enlightenment. Only when this idea was developed as one of the foundations of historicism by Herder, Möser, and the exponents of cultural and political pluralism and evolution, and propagated as a form of resistance to French ideas and imperialism, were its implications more fully absorbed in the military field.

With the great debates over French Revolutionary warfare, this idea came to the forefront of German military thought. It figured prominently in Scharnhorst's 'General Reasons for the French Success in the Wars of the Revolution' (1797), as well as in the works of Berenhorst and Bülow. Both the intellectual and the military environment of the young Clausewitz expounded this same idea.

Throughout his life, from his early studies of the Thirty Years War, this view of war within the context of its particular social and political reality was fundamental to Clausewitz's historical and theoretical outlook. It is also clearly revealed in his analysis of the great events of the wars of the Revolution and Napoleon, in which he followed in Scharnhorst's footsteps. Under Scharnhorst, he was one of the exponents of the military reforms which were closely linked to a comprehensive reform of Prussian society and politics, and which were based on a clear appreciation of the social and political sources of French power. In his review 'Prussia in her Great Catastrophe' written in the 1820s, Clausewitz elaborately expressed the assumptions that guided the reformers. In 1806 the army and administration were a product of the Frederickian absolutist state. While in the eighteenth century they had been perfected within the conditions and limitations of the *ancien régime*, they now became hopelessly inadequate in post-Revolutionary Europe.[1]

In *On War* Clausewitz outlined the transformation of warfare and the way it had been perceived:

> In the last decade of the eighteenth century, when that remarkable change in the art of war took place, when the best armies saw part of their doctrine become ineffective and military victories occurred on a scale that up to then had been inconceivable, it seemed that all mistakes had been military mistakes . . . [but] clearly the tremendous effects of the French Revolution abroad were caused not so much by new military methods and concepts as by radical

[1] 'Nachrichten über Preussen in seiner grossen Katastrophe', ch. 1, in Rothfels (ed.), *Schriften*, pp. 202–17.

changes in policies and administration, by the new character of government, altered conditions of the French people, and the like . . . Not until statesmen had at last perceived the nature of the forces that had emerged in France, and had grasped that new political conditions now obtained in Europe, could they foresee the broad effect all this would have on war.[2]

Not only military institutions and methods of warfare but also political aims and the conduct of operations are dependent upon the array of cultural, social, and personal circumstances. The growing realization of this was central to the transformation of 1827.

The Nature of War versus Policy: What were the Origins and Nature of Clausewitz's Dialectic?

As we might expect, a full understanding of the transformation in Clausewitz's thought in 1827, which resulted in the inclusion of the concept of limited war in military theory, cannot be gained without viewing the changes that occurred in his political perspectives. We saw that during the heroic period of the Napoleonic Wars and Prussia's struggle for independence, he tended to have in mind great and far-reaching political aims. This was the state of mind which guided his activities and harsh criticism of his country's policies, and which found clear theoretical expression in 1804 and 1812, when he twice outlined his outlook on war. This was also the state of mind wich favoured his conception of military decision.

However, with the end of the heroic period and with the return of the politics of European equilibrium, Clausewitz's concern shifted to the limited and complex considerations that these politics entailed. The problem that now claimed his attention was how to secure Prussia's status, strength, and stability within the European concert of powers against the dangers posed by both external and internal forces. This is what Paret calls the shift in Clausewitz's political outlook from an idealistic strain to an emphasis on *Ordnung*.[3] This perspective dominated his writings from that period on Prussia's foreign policy, and culminated in his works on European politics written in the last year of his life.

While it was not the direct cause of his conceptual change of course in 1827, the shift in Clausewitz's political state of mind provided a receptive background against which this change took root, and

[2] *On War*, VIII, 6B, p. 609.

[3] Paret, *Clausewitz*, p. 421.

in the process became itself conscious and pronounced. Here too, Clausewitz's political perspectives and military conceptions went hand in hand and complemented each other.

This twofold nature of the transformation in Clausewitz's thought, political and military, explains why in July 1827 he put forward two ideas as guide-lines for the revision of his work. The new idea that war can be of two types, aiming either at completely overthrowing the enemy or at a limited objective, appears first. This idea is explained by another: the character and scale of military operations are closely linked to the character and scope of the political objectives; consequently, great significance is now attached to the conception that war is but a continuation of policy by other means.[4] This conception might have been integral to Clausewitz's thought throughout his life; but when both policy and war had been viewed in expansive terms, it could not have had much significance.

As the depth of the crisis that Clausewitz's outlook on war and conception of theory underwent in 1827 has not been realized fully, the exact nature of his intellectual development during his last years has also remained somewhat vague. This has been particularly so since, in his attempt to resolve the crisis, Clausewitz borrowed from his cultural environment new intellectual devices whose origins and nature have also remained a mystery. For the sake of clarity, these developments will be treated separately. First, the nature of the transformation in Clausewitz's thought will be examined. Then, the new intellectual devices which he employed and which made possible his particular solution to the problem he faced will be traced and explained.

In brief, the late development of Clausewitz's thought can only be understood within the context of his attempt to bridge the gulf in his theory of war by reconciling his old conceptions of the nature of war which he did not abandon, with his new awareness of the diversity of wars in reality. As largely noticed by Aron, the revision in Clausewitz's thought took shape in two main stages. Beginning at the end of Book VI and continuing in Book VIII, the last book of *On War*, it was further developed in Book I, the only one that Clausewitz succeeded in addressing in his plan to revise the whole of the work.

[4] 'Note of 10 July 1827', *On War*, p. 69; see App.

At the end of Book VI Clausewitz realizes that the war of destruction is not the exclusive form of war, and that by ignoring that which does not conform to it, theory becomes cut off from historical reality. We have seen the devastating threat that this growing realization posed to his conception of the nature of war, which was dominated by the Napoleonic experience. What was now to become of this conception? Initially, Clausewitz was unprepared to abandon it. It was therefore necessary for him to devise an intellectual structure which would accommodate it together with his new ideas. He therefore recognizes the existence of two types of war, *but* claims that the war of destruction expresses the nature of war and thus takes priority; against half-hearted war, an all-out one would always gain the upper hand. A new concept now becomes necessary: 'the urge for decision' is 'true war, or *absolute war* if we may call it that'.[5] Limited wars are not a genuine form of war but the results of various factors which exercise counter-influences on the real, absolute nature of war and modify it.[6]

In Book VIII Clausewitz examines the problem extensively and compromises with the same solution:

> What exactly is this nonconducting medium, this barrier that prevents a full discharge? Why is it that the philosophical conception is not sufficient? [*der philosophischen Vorstellungsweise nicht Genüge?*] The barrier in question is the vast array of factors, forces and conditions in national affairs that are affected by war . . . Logic comes to a stop in this labyrinth . . . This inconsistency . . . is the reason why war turns into something quite different from what it should be according to its concept [*Begriff*] — turns into something incoherent and incomplete.[7]

However, since theory cannot ignore reality, one must leave

> room for every sort of extraneous matter. We must allow for natural inertia, for all the friction of its parts, for all the inconsistency, imprecision, and timidity of man; and finally we must face the fact that war and its forms result from ideas, emotions, and conditions prevailing at the time . . . Theory must concede all this; but it has the duty to give priority to the absolute form of war.[8]

Various factors which are *alien* to the nature of war therefore prevent it from fully realizing its true character. These factors are of two kinds.

[5] *On War*, VI, 28, pp. 488–9; my italics.
[6] Ibid. VI, 30, p. 501.
[7] Ibid. VIII, 2, pp. 579–80.
[8] Ibid. pp. 580–1.

First, within war itself factors of friction and uncertainty operate; and man himself, the actual agent of war, is a creature whose timidity and limited comprehension prevent him from fully carrying out the demands imposed by the nature of war on those who want to succeed. Clausewitz had already developed this conception in a work which was probably written during his period at Koblenz. 'On Progress and Stagnation in Military Activity'. Since the constitutive element of war was fighting, it was necessary to explain how there could be lulls or periods of low activity in war at all.[9] Now Clausewitz expands this explanation to include not only periods of limited activity within war but whole limited wars. And he adds a new component by claiming that apart from the internal interfering factors, war is also constrained by external forces. It does not exist in isolation, but is affected by the historical conditions out of which it arises. In most cases, war is not the dominant activity in the life of nations. A variety of other values, goals, and considerations guide nations and prevent a maximization of the conduct of war. All these factors, interior and exterior, are alien to the nature of war, but limit its intensity in practice. Limited wars, which include most of the wars in history, are therefore the result.

Hence the relationship between war and politics, which encompasses most of the exterior factors mentioned above. In Book VIII, 'War Plans', Clausewitz expounds upon the full implications of his new ideas, asserting that the scale, character, and objectives of the military operations result largely from an interplay with the scope and nature of the political aims. The explication of this point in particular was an original contribution of Clausewitz, to be further developed only with the modern study of international relations. The influence of the political aim on the objective of operations, he wrote, 'will set its [the war's] course, prescribe the scale of means and effort which is required, and make its influence felt throughout down to the smallest operational detail'.[10] He elaborates on this point in the chapter on the 'Scale of the Military Objective and the Effort to be Made'. Both, he claims, are governed by 'the scale of political demands on either side', as well as by the characteristics of the belligerents and by their reciprocal actions which lead to the escalation

[9] 'Über das Fortschreiten und den Stillstand der Kriegerischen Begebenheiten', in *Zeitschrift für Preussische Geschichte und Landeskunde*, XV (1878), 233–40. The main ideas of this work appear in the early part of *On War* as ch. 16 of Bk. III.
[10] *On War*, VIII, 2, 579.

of the conflict.[11] This is also the theme of the chapter on 'The Effect of the Political Aim on the Military Objective'. He concludes that 'once this influence of the political objective on war is admitted, as it must be, there is no stopping it; consequently we must also be willing to wage such minimal wars which consists in *merely threatening the enemy*, with *negotiations held in reserve*'.[12] Finally, in the celebrated chapter entitled 'War Is an Instrument of Policy' Clausewitz fully elaborates the idea that war cannot be understood outside the political context:

> War is only a branch of political activity . . . it is in no sense autonomous . . . The main lines along which military events progress, and to which they are restricted, are political lines that continue throughout the war into the subsequent peace . . . All the factors that go to make up war and determine its salient features—the strength and allies of each antagonist, the character of the people and their governments, and so forth . . . are these not all political?[13]

However, these widely quoted passages form only half of the picture. While all the characteristics of war are decisively influenced by politics, this influence is by no means part of the nature of war; on the contrary, the influence of politics is an *external* force which works *against* the true essence of war, harnesses it to its needs, and in the process modifies the imperatives which it imposes. 'In making use of war, policy evades all rigorous conclusions proceeding from the nature of war . . . [It] converts the overwhelmingly destructive element of war into a mere instrument.'[14]

The historical survey of the development of the art of war—to which we have already referred several times—was in fact far from being a detached, disinterested, historicist study.[15] We are now in a position to see its actual purpose. The survey was part of the process by which Clausewitz laboured to clarify his thoughts and aimed to examine concretely (*a*) the array of conditions which had prevented the realization of the true, absolute nature of war in most periods of history; (*b*) the circumstances in which this nature had appeared under the Romans, Alexander the Great and, obviously, the French Revolutionaries and Napoleon; and (*c*) the resulting theoretical conclusions.

[11] Ibid. VIII, 3B, p. 585
[12] Ibid. VIII, 6A, pp. 603–4.
[13] Ibid. VIII, 6B, pp. 605–6.
[14] Ibid.
[15] As claimed by Paret, *Clausewitz*, pp. 348–9.

Under the *condottieri*, 'war lost many of its risks; its character was wholly changed, and no deduction from its proper nature was still applicable'. War was also limited during the *ancien régime*. 'All Europe rejoiced at this development. It was seen as a logical outcome of Enlightenment. This was a misconception. Enlightenment can never lead to inconsistency . . . [Indeed] so long as this was the general style of warfare with its violence limited in such strict and obvious ways, no one saw any inconsistency in it.' But the Revolution and Napoleon unleashed the forces contained in society, and war then 'took on an entirely different character, or rather closely approached its true character, its absolute perfection . . . untrammelled by any conventional restraints, [it] had broken loose in all its elemental fury'. What does the future hold? Will limited wars reappear? This, Clausewitz wrote, was difficult to answer, yet limited wars were not very likely in the future: 'once barriers—which in a sense consist only in man's ignorance of what is possible—are torn down, they are not so easily set up again.'[16]

What is the theoretical conclusion of all this?

> We can thus only say that the aims a belligerent adopts, and the resources he employs, must be governed by the particular characteristics of his own position; but they will also conform to the spirit of the age and to its general character. Finally, they *must also be drawn from the nature of war*.[17]

The compromise that Clausewitz worked out between his lifelong view of war and his new awareness that the conduct of war takes many forms and that this is so primarily because of changing political conditions, led him, therefore to develop a new theory which recognized two types of war, but regarded the one, absolute war, to be the genuine expression of the nature of war, and superior to the other. Yet, as he continued to probe his new ideas, this theory became insufficient. The chapter 'War is an Instrument of Policy' marked a further shift. If the understanding of war was dominated by its political function, the primacy given to absolute war lost much of its point. In the dilemma between his lifelong view of war on the one hand, and the diversity of political aims and military operations in historical reality on the other, Clausewitz was moving a further step towards the latter. Nevertheless, he did not altogether abandon

[16] *On War*, VIII, 3B, pp. 587, 591, 593.
[17] Ibid. VIII, 3, p. 594; Clausewitz's emphasis.

the core of his old conception, that the constitutive element of war, fighting, dominates the nature of war. Nor did he relinquish his belief in the superiority of the engagement and the clash of forces over all other military means. This was the basis for the amended compromise of Book I, which Clausewitz revised, as he had planned in July 1827, after he completed Book VIII, the last book of *On War*.[18]

In Book I, the essence of war is also presented as an eruption of force and violence: 'The impulse to destroy the enemy . . . is central to the very concept [*Begriff*] of war . . . war is an act of force, and there is no . . . limit to the application of that force.'[19] However, the unlimited nature of violence in war no longer relies directly on the notion that all-out war is clearly superior; Napoleonic warfare is no longer perceived as the only correct form of war. Violence in war is now presented in connection with tendencies towards escalation which are inherent in the interplay between the belligerents' aims and efforts.

Clausewitz's extensive argumentation can be summarized as follows: the nature of war implies that the aim of the belligerents must be the total destruction of the enemy's ability to fight, because otherwise his complete surrender will never be secured. In addition, since each side attempts to surpass the other's efforts, escalation and a tendency to maximize the mobilization of forces is also created.[20] Thus, 'were it a complete, untrammelled, absolute manifestation of violence (as the pure concept would require) war would of its own independent will . . . rule by the laws of its own nature, very much like a mine that can explode only in the manner or direction predetermined by the setting.'[21]

Why then, if war is 'pulsation of violence', does it not discharge itself in a single explosion? Why is it divided into several engagements, and why does it last for long periods of time, occasionally lacking energy and determination, and even falling into inactivity?[22] Clausewitz's reply in Book I elaborates the argument put forward in Book VIII (itself, as mentioned, an expansion of an early work) with one significant change; the lulls in military activity are no longer seen in a negative light, and the factors which explain

[18] See App.

[19] *On War*, I, 1, s.3, pp. 76–7.

[20] Ibid. I, 1, s. 3–5, p. 77.

[21] Ibid. I, 1, s.23, p. 87.

[22] Ibid. I, 1, pp. 79–80; s.12–19, pp. 81–5; the quotation is from s.23, p. 87.

them no longer include man's imperfection and timidity. War is not discharged in a single explosion due to the activity of various factors within war which are summarized for the most part by the concept of friction, and owing to influences, mainly political, which are exterior to war.

It is again important to understand that the influences of politics on war do not belong to the nature of war, but, on the contrary, contradict it. The political influences 'are the forces that give rise to war; the same forces circumscribe and moderate it. They themselves, however, are not part of war . . . To introduce the principle of moderation into the theory of war itself would always lead to . . . absurdity.'[23] Politics thus places itself above war and modifies it to suit its needs.

The modifications of the nature of war by the actual context in which war takes place, therefore completely change its character. At this point Clausewitz announces the opening of an entirely new discussion concentrating on the characteristics of war in reality.[24] The aims and means of war are no longer dictated by the maximal imperative inherent in the nature of war, but vary according to each particular case. The aim of war is shifted from the total overthrow of the enemy to the aim put forward by politics. Consequently, war is no longer conducted on a total scale but according to the requirements of the political aim. Clausewitz again discusses in depth the interaction between the scope of the political aim, its importance to the parties involved, and the scale of the effort required to achieve it.[25]

As indicated in the opening passage of Book I, chapter 2, Clausewitz's revision of his military theory is now applied to a closer examination of the purpose and means in war. This explication is highly significant because it clearly reveals how far Clausewitz had retracted from his belief in all-out decision. It can be summarized as follows: he now gives an equal status to a variety of war aims and operational objectives, or, to use the terminology of 1804, purposes of war and purposes in war. But he still regards the clash of forces as the dominant means for the attainment of any purpose, and treats with suspicion any means to evade it.

Following the arguments of chapter 1,

[23] *On War*, I, 1, s.3, p. 76.
[24] Ibid. I, 1, s.6, p. 78.
[25] Ibid. I, 1, s.10–11, pp. 80–1.

we can now see that in war many roads lead to success, and that they do not all involve the opponent's outright defeat. They range from *the destruction of the enemy's forces, the conquest of his territory, to a temporary occupation or invasion, to projects with an immediate political purpose, and finally to passively awaiting the enemy's attacks.*[26]

However, this plurality does not extend to the military means. Already in chapter 1, echoing his criticism of Bülow in 1805, Clausewitz had written: 'Kind-hearted people might of course think there was some ingenious way to disarm or defeat an enemy without too much bloodshed, and might imagine this is the true goal of the art of war. Pleasant as it sounds, it is a fallacy that must be exposed.'[27] Now, in chapter 2, Clausewitz wrote on the means in conducting war, reiterating his positions from 1804: 'There is only one: *combat*. However many forms combat takes . . . it is inherent in the very concept of war.'[28] The decisive clash of forces may be supported by other means. It may not even take place at all but still exert decisive influence merely by its probability and expected outcome. Yet, in any case, the 'destruction of the enemy forces is always the superior, more effective means, with which others cannot compete'.[29]

To conclude,

our discussion has shown that while in war many different roads can lead to the goal, to the attainment of the political object, fighting is the only possible means. Everything is governed by a supreme law, *the decision by force of arms* . . . A commander who prefers another strategy must first be sure that his opponent . . . will not appeal to that supreme tribunal . . . If the political aims are small, the motives slight and tensions low, a prudent general may look for any way to avoid major crises and decisive actions . . . and finally reach a peaceful settlement. If his assumptions are sound and promise success we are not entitled to criticize him. But he must never forget that he is moving on devious paths where the god of war may catch him unaware.[30]

The men of the nineteenth century, criticized for tendentious interpretation, were therefore perfectly justified here in regarding Clausewitz's writings as the classic formulation of their belief in the dominance of the great battle. On this point Clausewitz held effectively the same position throughout his life. Indeed here too,

[26] Ibid. I, 2, p. 94. [27] Ibid. I, 1, s.3, p. 75.
[28] Ibid. I, 2, p. 95. [29] Ibid. p. 97. [30] Ibid. p. 99.

it is in our period that Clausewitz's position has tended to be interpreted in terms that conveniently correspond to contemporary views of war.[31] This endemic misinterpretation of Clausewitz's ideas has been mainly due to the failure to grasp fully the origins and nature of the transformation of his thought and, particularly of the new intellectual forms in which this transformation was expressed.

It is not surprising that since publication, *On War* has had the reputation of being a very difficult and complicated philosophical work. The interested reader wishing to read the treatise which, from the time of the German wars of unification and the domination of the German military school was considered to be the masterpiece of military thought, encounters in the first and basic chapter of the book a highly complex intellectual structure, which hardly reveals a 'commonsense' understanding of war. He reads about 'absolute war' that was first defined in maximal and dramatic terms but was then totally overturned and assumed completely different expressions in reality as an instrument of policy. He has no means of understanding the nature and origins of this structure which was supported by an equally puzzling argument that explained why any limitation was alien to the nature of war, and elaborated the reasons for the lull in military activities and for their duration over substantial periods of time. Since its appearance, *On War* was therefore known for being much quoted but little read.

Ironically, this situation may have even enhanced Clausewitz's reputation. The men of the nineteenth century, in any case, emphasized most of the same points as Clausewitz's, and the obscure and elaborate reasoning of Books I and VIII only added to Clausewitz's image of profundity, as they were regarded as demonstrating the 'philosophical' manner of expression that was only to be expected of a philosophical masterpiece on war. It was generally

[31] Paret, for instance, writes: 'Clausewitz's supposed preference for the major, decisive battle, in particular, is an erroneous assumption, based on the very inability to follow his dialectic that he had predicted.' (Paret, *Clausewitz*, p. 369.) As evidence, Paret cites the passage on the variety of roads leading to success in war, but fails to cite Clausewitz's further emphasis that the pluralism of objectives is not matched by a pluralism of means. 'Clausewitz's dialectic' and the inability to follow it are discussed below. Here it is merely necessary to clarify that Clausewitz, of course, did not mention dialectic or anything to that effect when expressing apprehension that his work would be misunderstood (indeed, he did not mention dialectic anywhere).

assumed that this manner of reasoning was somehow related to the highly influential German idealistic philosophy (famous, or infamous, for the difficulties in understanding it) and especially to Hegel. Shortly after the publication of *On War* one critic, alluding precisely to this view, wrote:

> The streams whose crystal floods pour over nuggets of pure gold do not flow in any flat and accessible river bed but in a narrow rocky valley surrounded by gigantic Ideas, and over its entrance the mighty Spirit stands guard like a cherub with his sword, turning back all who expect to be admitted at the usual price for a play of ideas.[32]

Camon conveyed the same impression though from a point of view which was much less favourable than that of most of his contemporaries. In a much quoted passage he described Clausewitz as: 'The most German of Germans . . . In reading him one constantly has the feeling of being in a metaphysical fog.'[33] This was the closest one could get to admitting a failure to understand what Clausewitz actually had in mind.

Unfortunately, Clausewitz's modern interpreters too have been puzzled by his late intellectual formulations. We now have the advantage of possessing a sequence of Clausewitz's early works which provide an almost continuous picture of the development of his thought from 1804. Equally helpful is the fact that Clausewitz did not live to finish the revision of *On War*, and that the book we possess is therefore a draft that provides a history of the course of the work, almost linearly documenting the development of his thought, the problems he encountered, and his attempts to resolve them. Yet, the objective difficulties of the subject and biased approaches to it have reinforced each other in obscuring the nature and development of Clausewitz's ideas.

The interpretations of chapter 1 of Book I, which represent hardly more than an attempt to paraphrase Clausewitz's own words, have reflected this chronic confusion. In the struggle to understand his ideas it has often proved difficult to see the wood for the trees. Aron,

[32] *Preussische Militair-Literatur Zeitung*, 1832, quoted by Howard, 'The Influence of Clausewitz', in *On War*, p. 27.

[33] Camon, *Clausewitz*, p. viii. Bernard Brodie, one of the chief contributors to the 'Clausewitz renaissance', dismissed this complaint with the words: 'This is simply nonsense'. This remark is, however, an unfortunate reflection on Brodie's own 'high-spirited' commentary to *On War*: 'The Continuing Relevance of *On War*' and 'A Guide to the Reading of *On War*', in *On War*, the quotation is from p. 18.

whose interpretation of Clausewitz's major theses is the most comprehensive and penetrating, has unfortunately only perfected this tendency by attempting to place Clausewitz's formulations in chapter 1 in some meaningful general context. According to Aron, Clausewitz first uses an 'abstract model' which exists only in 'the world of concepts and ideals', and then shows how this model operates in reality.[34] Now firstly, Clausewitz never believed in a 'world of concepts and ideals',[35] But, apart from that, why did Clausewitz need this kind of 'abstract model' at all? According to Aron, chapter 1 is simply the culmination of 'Clausewitz's system', which has always first distinguished sharply between concepts by way of 'antithesis' and then examined their actual appearance in reality.[36] Following in Schering's footsteps Aron, therefore, interprets the whole of Clausewitz's thought on the basis of the 'dialectic' between ends and means, moral and physical, defence and attack, and even number and morale, boldness and caution, and ambition and risk.[37]

The passage cited by Aron as revealing 'Clausewitz's method throughout his life' is taken from Clausewitz's critique of Bülow written in 1805. There, Clausewitz criticizes Bülow's definitions of strategy and tactics for not being clear and for overlapping each other. To define distinct concepts, Clausewitz writes, their boundaries must be delineated precisely. For this, the nature of the concept in question should be traced until it reveals a change at its extreme limit. This point is the concept's boundary.

A certain similarity in sound has led Aron to interpret an entirely different matter in the spirit of 1827–30.[38] The passage of 1805 is a lesson in clear and distinct definitions which looks like a typical product of the logic lessons that Clausewitz had just attended at the Institute, and which very probably reflected the influence of Kant, the great distinction-maker, through the medium of Kiesewetter.[39]

[34] Aron, *Clausewitz*, pp. 106, 114.

[35] In this respect a trans. of Clausewitz's *Abstraction* and *Wirklichkeit* (*On War*, I, 1, p. 6), or *wirkliche Welt* and *blosse Beggriff* (*On War*, I, 1, p. 8) as 'real world' and 'abstract world' may be misleading. Clausewitz never speaks of an abstract world or a world of ideas.

[36] Aron, *Clausewitz*, pp. 104, 111.

[37] Ibid. 322, 325.

[38] 'Bülow', in Hahliveg (ed.), *Verstreute kleine Schriften*, p. 68. It must be noted that the word 'extreme' (*extrem*) used in 1805, does not appear at all in Bk. I, ch. 1. Again the English trans. might be misleading here. In describing the tendency of war to magnify, Clausewitz consistently uses the term *Äusserst*.

[39] As suggested by Gallie; see below, n. 42.

It is by no means said there that the nature of the concept lies at its extreme; on the contrary, the concept's extreme boundaries are revealed by the very fact that henceforth, by definition, the essence of the concept ceases to be in force. Throughout his life Clausewitz defined his subject-matter clearly and distinctively and never saw the essence of a phenomenon in its most extreme expression. That fighting, the constitutive element of war, should be interpreted in expansive terms stems, as we have seen, not from Clausewitz's 'logical method' but from his lifelong outlook on war based on the dominating experience of his age. In his attempt to gain a coherent understanding of the mystery of Clausewitz's later formulations, Aron has created a myth of 'Clausewitz's lifelong method'.

The myth is in fact revealed by Aron's own argument. 'All his life,' he wrote, 'Clausewitz practised the method put forward in 1805, or rather half of this method; namely he chose, as the point of departure, extremes or complete antitheses. There is hardly a trace of the search for boundaries in the Treatise [*On War*].'[40] Indeed, there is no search for boundaries in *On War*, but this is precisely the issue in 1805; there may be antitheses in the latter parts of *On War*, but there are none in 1805! As an example of the 'system of antitheses' which is supposed to have characterized Clausewitz's writing throughout his life, Aron turns to the end of Book VI of *On War* (written in 1826–7).[41] This is no coincidence; no earlier example exists. In all his works prior to *On War* and in most of *On War* itself, nothing in Clausewitz's writing, generally characterized by its clarity and realistic approach, comes close to the formulations of his last years which gave his work the reputation of being covered by 'metaphysical fog'. This should have been obvious. Though largely aware of the transformation of Clausewitz's military outlook, Aron failed to realize its scope and implications for Clausewitz's theoretical approach. While he noted the late appearance of the concept of 'absolute war' and its close link to Clausewitz's late development, he sought to explain it by referring to an early 'method'. Instead of clarifying Clausewitz's late ideas, he obscured his earlier ones as well.

The fact that something was very wrong with Clausewitz's reasoning in Books VIII and I, and consequently also in the way it was usually interpreted was finally noticed by W. B. Gallie. Clausewitz's interpreters, he wrote, struggled in vain to explain his

[40] Aron, *Clausewitz*, p. 79. [41] Ibid.

intellectual structure, mistakenly assuming that this structure was coherent. Not being a specialist on Clausewitz, Gallie himself failed to reveal the historical and intellectual origins of Clausewitz's problematic formulations. He was not aware of the development of Clausewitz's ideas, accepted Aron's conception of 'Clausewitz's lifelong method' and merely sought to correct its 'logic'. Yet, Gallie at last exposed the fact that had embarrassed Clausewitz's interpreters: Clausewitz's conceptions, he maintained, were the result of a tension which could not be reconciled between his definition of war itself and his notion that war was a political means.[42]

A full understanding of the theoretical formulations that have created so much confusion can only be achieved by realizing the interaction between the theoretical crisis in which Clausewitz found himself in 1827 and the intellectual devices that his cultural environment offered him at that same time. The preservation of the core of his old conceptions within his new ones, despite their contradictory nature, was made possible, and even perceived by Clausewitz as an achievement, by borrowing from the most ambitious intellectual attempt at an all-encompassing and integrative explanation of all the contrasts and contradictions of reality; namely, the German idealistic philosophy which was elevated by Hegel at precisely that time to a zenith of power, and whose influence on Clausewitz has always been the subject of wonder and speculation.

In a cautious attempt to delineate the affinity of Clausewitz's thought to German idealism, Paret stressed in particular that from the intellectual climate of the period, Clausewitz absorbed the emphasis of idealism on the integrative interrelation of all phenomena as well as a tendency to a dialectic discussion in terms of theses and antitheses, contradictions, polarity, activity and passivity, positive and negative.[43] In fact, Paret too projected the image of Clausewitz's late work on his earlier writings. From his youth until the final stages of his work on *On War*, Clausewitz shows no substantial affinity to the distinctive doctrines of idealism; but he does reveal the decisive influence of these doctrines during his last years.

First, the fact that has somehow been lost sight of must be stressed again; in all of Clausewitz's extensive writings until the last stage of

[42] W. B. Gallie, 'Clausewitz On the Nature of War', *Philosophers of Peace and War, Kant, Clausewitz, Marx, Engels and Tolstoy* (Cambridge, 1978), esp. pp. 48–65.

[43] Paret, *Clausewitz*, esp. pp. 5, 85, 150–1.

his life, there are no theses and antitheses, no polarity or dialectic (unless of course one reads them into ordinary reasoning and simple contrasts and reciprocal relations) nor, indeed, any mention of 'absolute war'. Nor do they appear in the early or unrevised books of *On War*, which continue Clausewitz's lifelong train of thought. As for the quest for an all-encompassing and comprehensive explanation of reality, this had been one of the principal themes of the German Movement as a whole from the days of the 'Storm and Stress'. While Clausewitz continued this quest throughout his life, only in the last phase of his work did it assume the totally integrative character unique to idealism.

Clausewitz's world-view and intellectual affinities should also be understood from another perspective and from a psychological point of view. As pointed out by Paret, Clausewitz was not a professional philosopher but a typical educated representative of his period who absorbed attitudes and scraps of ideas, not necessarily at first hand, from his cultural environment.[44] To this, however, it should be added that unlike any typically educated person of his period, Clausewitz was throughout his life motivated by the desire to work out a comprehensive view of war, and naturally he was highly sensitive to anything in his cultural environment which could have had a bearing on the realization of this aim. This kind of involvement and interest partly explains the fact that Clausewitz drew mainly on what had already been considered classic literature: Machiavelli, Montesquieu, the great figures of the 'Storm and Stress' movement and German *Klassizismus*, Kant, and so on. By contrast, Fichte's or Schelling's idealism (as, for that matter, Romanticism) was in the first decade of the nineteenth century a radical trend, albeit of wide-ranging reputation. As we have seen, Clausewitz shared many of the ideas of the Romantics but was far from agreeing with their overall outlook. The same applied to idealism. In a letter to his fiancée on 15 April 1808, Clausewitz refers to one of Fichte's political works: 'he has a manner of reasoning that pleases me very much, and I felt that all my tendency to speculative reasoning was awakened and stimulated again'.[45] Later, following Fichte's article on Machiavelli, Clausewitz even wrote directly to the famous philosopher. However, there is no sign that he was effectively influenced by Fichte's philosophy or dialectic. On the contrary, while sharing Fichte's

[44] Ibid. 151. [45] Schwartz, *Leben*, i. 305.

patriotic sentiments and emphases on moral forces and creativity, Clausewitz, as noted by Paret, was clearly a 'realist' and rejected purely spiritual entities, any form of 'mysticism' and teleological conceptions of history.[46] In short, he shared no affinity with idealistic metaphysics. He aimed at a realistic military theory, firmly grounded in historical experience and in the nature of war.

However, by 1826–7 both Clausewitz's theoretical expectations and the status and power of the idealistic philosophy had changed drastically and their paths had crossed. It became clear to Clausewitz that there was a serious discrepancy between his conception of the universal nature of war and the test of historical experience. While regarding both to be indispensable, he was forced to reject one of them. Fortunately, in the same years in Berlin, Clausewitz's city of residence, Hegel's idealism was reaching a climax of influence, unequalled in Germany since the days of Kant. And one of the chief lessons of this philosophy was that all the contrasts and contradictions of reality were actually but differing aspects of a single unity. In this, the 'identity' ideal of all phenomena inherent in the German Movement was brought to its pinnacle. Clausewitz was, therefore, not compelled to resolve the contradiction created in his mind by abandoning one of the two conceptions that he regarded as essential; on the contrary, resolving this contradiction, while keeping its components by viewing them from a higher standpoint, was now perceived as an achievement and an indication that his theory of war was on the right systematical road.

Was Clausewitz then a disciple of Hegel, and if so, how was he influenced? This question has been the cause of much speculation since the publication of *On War*, repeatedly expressed by as different and remote commentators as the above-mentioned Prussian military critic of 1832 and Lenin.[47] The first attempt to tackle it was made in 1911 by Lieutenant-Colonel Paul Creuzinger in *Hegel's influence on Clausewitz*.[48] If we are to believe Creuzinger, there is not a single idea in *On War*, from tactical conceptions to strategical outlook, that is not shaped by Hegel's influence. Creuzinger knew that in all probability Clausewitz could only have been influenced

[46] Paret, *Clausewitz*, pp. 151, 350; and p. 183 of this work.

[47] V. I. Lenin, 'The Collapse of the Second International', *Collected Works* (Moscow, 1964), 21. 219.

[48] Paul Creuzinger, *Hegels Einfluss auf Clausewitz* (Berlin, 1911). Lenin possibly relied on this work in his explicit presentation of Clausewitz as Hegel's disciple.

by Hegel from the 1820s, for prior to that, Hegel had been relatively unknown. But as Creuzinger was only familiar with *On War*, he was unaware of the fact that most of the conceptions that he attributed to Hegel's influence had already been outlined by Clausewitz in his early works.

Unfortunately, Creuzinger's work placed the whole argument on a totally misleading course. In reaction to Creuzinger, Schering laboured to show that Clausewitz's supposed dialectic was not exactly similar to Hegel's.[49] In Schering's footsteps went both Paret, who added that Clausewitz's dialectic could have been influenced by many others apart from Hegel,[50] and Aron, to whom Schering's argument appeared particularly valid in view of the abundance of 'antitheses' and dialectic relationships he found in Clausewitz's work. Aron also went to great lengths to show that Clausewitz's conceptions had no affinity to Hegel's metaphysics.[51] While Schering, Paret, and Aron did not totally rule out the possibility that Hegel might have somewhat influenced Clausewitz, they dismissed this possibility almost completely and conferred upon it (in view of Creuzinger's assertions, justifiably) a dubious image.

What, then, do we know about Clausewitz's affinity to Hegel? In contrast to Fichte's case, we have no reference to Hegel in Clausewitz's writings. Yet this does not mean much; in Clausewitz's letters to Marie, the main source for his biography, there is a large gap in the 1820s when they lived together in Berlin; and these were precisely the years when Hegel served as rector of the University of Berlin and his reputation achieved unprecedented heights. Indeed, we do possess contemporary evidence, revealed by Paret, which almost certainly proves that Clausewitz was acquainted with Hegel in the salons of Berlin.[52]

As for the influence of Hegel's ideas, we do not know whether, and how much, Clausewitz read Hegel or indeed understood him, or, alternatively, whether he absorbed some of Hegel's ideas from the intellectual environment in Berlin. However, all that we do know of Clausewitz's intellectual interests and involvement makes it highly improbable that the philosophy which achieved such widespread

[49] Schering, *Kriegsphilosophie*, pp. 111–19.
[50] Paret, *Clausewitz*, pp. 84, 150.
[51] Aron, *Clausewitz*, pp. 321–31.
[52] H. Hoffman von Fallersleben, *Mein Leben* (Hanover, 1868) i. 311–12; cited by Paret, *Clausewitz*, p. 316.

influence failed to attract his attention. And, above all, we have the highly distinctive, new intellectual patterns in his late work to support this.

Indeed, this work does not reveal any affinity to Hegel's metaphysics, idealism, or conception of history. But it does reveal what appears to be a direct influence of Hegel's political and social ideas, which will be discussed in the next section. Furthermore, it reveals a new and vigorous use of dialectic tools, along with a much stronger comprehensive and integrative ideal. The question as to whether this new dialectic was exactly like Hegel's, or the argument that from his youth Clausewitz had come in contact with the dialectics of Fichte, Schleiermacher and, perhaps, Schelling, and the all-embracing 'identity' quest of the German Movement, miss the point. Clausewitz adapted scraps of ideas to his needs, and his distinctive use of dialectic tools together with a new forceful emphasis on the totally integrative nature of theory only made an appearance in the later stages of his work, during the period in which idealism and Hegel's influence surged to a peak.

The integrative quest of the period is forcefully revealed in Clausewitz's early treatise on war in 1816–18, where he betrays a certain fear that his work, intelligent as it might be, lacks the real internal, unifying logic to be *the* desired 'Theory of War'. 'Perhaps a greater mind,' he wrote, 'will soon appear to replace these individual nuggets with a single whole cast of solid metal, free from all impurity.'[53] In this respect the transition from Book VI, 'Defence', to Book VII, 'The Attack', marked a turning-point, apparently brought about by two discussions in which Clausewitz was then engaged. The first was the interesting interrelationship between defence and attack, already vaguely emerging in the *Principles of War for the Crown Prince* (1812), but extensively developed in Book VI.[54] Elaborating on this, Clausewitz appears to have come to the view that this interrelationship could perhaps be given a tighter theoretical expression. Precisely then, at the end of Book VI, the problem of the two types of war and the discrepancy between the nature of war and historical experience was added. Both issues now invited the employment of a new and highly acclaimed intellectual device: dialectic reconciliation.

[53] 'Author's Preface', *On War*, pp. 61–2.

[54] See p. 207, n. 26.

In Book VII, on attack, Clausewitz's attraction to this new device is still only alluded to, but unmistakably so. The book opens with a chapter on the relationship between attack and defence:

Where two concepts [*Begriffe*] form true logical contrasts [*Gegensätze*],[55] each complementary to the other, then fundamentally each is implied in the other. The limitation of our mind may not allow us to comprehend both simultaneously, and to discover by contrast the totality [*Totalität*] of one in the totality of the other. Nevertheless each will shed enough reciprocal light to clarify many of the details.[56]

This distinctive formulation, hitherto unprecedented in his writings, strikingly shows that the dialectical reasoning which was becoming dominant in Clausewitz's intellectual environment by the mid-1820s, influenced his own thought decisively. While apart from this opening statement Clausewitz hardly employed dialectic in Book VII, he used it with increasing skill and in a highly significant role in Book VIII and in the revision of Book I.

In the famous chapter 'War Is an Instrument of Policy', Clausewitz finally resolves the contradiction in his mind between war as the all-out use of force and the varying degrees of limited war revealed in historical experience, without relinquishing either of these ideas. War as a political and multi-faceted phenomenon is the unity that fuses the pure nature of war, which constitutes merely a partial understanding of reality, with the political conditions and requirements:

Up to now, we have considered the difference that distinguishes the nature of war [*Natur des Krieges*] from every other human interest, individual or social . . . We have examined this incompatibility from various angles so that none of its conflicting elements should be missed. Now we must seek out the unity into which these contradictory elements combine in real life, which they do by partly neutralizing one another . . . Being incomplete and self-contradictory it [war] cannot follow its own laws, but has to be treated as a part of some other whole; the name of which is policy . . . Thus the contradictions in which war involves . . . man, are resolved . . . Only if war is looked at in this way does its unity reappear; only then can we see that all wars are things of the same nature.[57]

[55] Since here, as well as in Bks. VIII and I, Clausewitz never used the concept *Idee*, it would perhaps be preferable not to translate *Begriff* and *Gegensätze* as idea and antithesis which tend to assume the required.

[56] *On War*, VII, 1, p. 523.

[57] Ibid. VIII, 6B, pp. 605–6.

The unity of the phenomenon of war, that is, the constitutive element common to all wars, is salvaged. The 'primordial violence, hatred, and enmity' of the nature of war are directed by the 'commander's creative spirit' through the 'play of chance and probability' to achieve the political aim. This is the 'remarkable trinity' which is presented by Clausewitz at the end of the first chapter of Book I, and which makes war 'more than a true chameleon that slightly adapts its characteristics to the given case'.[58]

Indeed, in the first chapters of Book I, Clausewitz's dialectic reaches its peak, and his conception of the nature of war finds its place in the actual diversity of war which previously threatened to invalidate it. Clausewitz has not become an idealist nor does he believe in any 'world of ideas'. He considered the concept of absolute war as an analysis of the actual forces which in his view comprise the nature of war. It was possible for him to maintain this view by claiming that this nature never existed in isolation, but always interacted with the other forces and influences of reality, chiefly politics, which modified and governed its original tendencies. A new intellectual tool assisted him in devising what he regarded as an adequate solution to the crisis into which his universal theory of war had fallen in 1827.

Political and Ethical World-View

The idea that the military command had to be subordinated to the political leadership was regarded by Clausewitz as a direct implication of the close link between the conduct of war and political aims. This idea stemmed, however, from much deeper historical and intellectual origins, and reflected Clausewitz's political and ethical outlook.

The accepted view that Clausewitz refrained from dealing with the ethical aspects of war, and that he confined himself to the study of war 'as it is',[59] requires careful historical scrutiny, though based on an apparently unequivocal statement by Clausewitz himself. After describing the advantages of guerrilla warfare he wrote: 'the question only remains whether mankind at large will gain by this further expansion of the element of war; a question to which the answer should be the same as to the question of war itself. We shall leave

[58] *On War*, I, 1, 28, p. 89.

[59] See e.g. Werner Hahlweg, *Carl von Clausewitz* (Göttingen, 1957), 62; cited with approval by Paret, *Clausewitz*, p. 352; see also pp. 348–9.

both to the philosophers.'[60] This statement also corresponds with Clausewitz's general tendency to avoid too direct a reference to philosophical questions about which he did not feel professionally qualified, and which might expose his work to criticism outside the military sphere.

However, to deduce from this that Clausewitz had no views on ethics in the framework of his general world-view, or that his outlook on war was divorced from this world-view seems inconceivable, particularly as we are dealing with a man for whom a comprehensive understanding of reality was a genuine need and the object of continuous efforts, who had an acute historical sense, and whose life was marked by a deep political commitment expressed in highly charged statements. Paret emphasizes that Clausewitz's historicist approach rendered his historical outlook almost totally free from value-judgements which assume universal, supra historical standards of measurements. This is reinforced, according to Paret, by Clausewitz's special point of view; he avoided ideological positions because his concern was with 'the diplomatic and military efficiency of any political community', expressed in 'results, which are judged in terms of energy and force'.[61] Though Paret's work is the most extensive study of Clausewitz's affinity to the state, this interpretation in fact totally obscures the real context of that affinity and its implications for the understanding of Clausewitz's world-view as a whole. 'Which Side was Clausewitz On?' asked C. B. A. Behrens in a concise and penetrating review of Paret's book, undermining the almost liberal image that Clausewitz has acquired in the West of today.[62] Indeed, 'military and political efficiency of political communities judged in terms of energy and force' is not a valueless standard of measurement; rather, it is a striking expression of Clausewitz's political and ethical outlook, deeply embedded in his intellectual milieu.

Here too, Clausewitz was a true child of his time. He operated during the fateful transition of German national consciousness from its Enlightenment, eighteenth-century forms, as expressed either in the humanitarian, cosmopolitan, and cultural orientation of, for example, Kant, Herder, Möser, and Schiller, or in the strict and limited framework of the absolutist state. These forms were transformed radically with the French imperialist threat, the

[60] *On War*, VI, 26, p. 479.

[61] Paret, *Clausewitz*, pp. 348, 352.

[62] C. B. A. Behrens, 'Which Side was Clausewitz On?' in *The New York Review of Books*, 14 Oct. 1976.

humiliating defeats at the hands of Napoleon, and the political settlements that he imposed on Germany. The fervent awakening of the German national movement which resulted, swept throughout society in a highly political form and with a strong emphasis on the dominant role of the state. These trends found expression in the deeds and the legend of the German war of independence of 1813, and were formulated into what was to be called the German conception of the state. This conception was carried on by the German historical school, strengthened by the failure of the liberal vision in 1848, and sanctified after the establishment of the German Reich by Bismarck's ingenious *Realpolitik*.[63]

The intention here is obviously not to understand Clausewitz in Treitschke's terms, or against the background of the height of militarism and social Darwinism in Wilhelmine Germany. However, Clausewitz's political and ethical outlook, and thus also his view of war, cannot be understood without realizing his position during the crisis of Prussian absolutism and at the formative period of a famous and highly influential intellectual tradition that gave Germany its unique place, separate from the political philosophy of the liberal West.

'German thought' wrote Troeltsch, 'whether in politics or in history or in ethics, is based on the ideas of the Romantic Counter-Revolution.'[64] Clausewitz operated in the historical and intellectual environment that, among others, gave rise to Fichte, Adam Müller, Savigny, and Hegel, all of whom, incidently, he probably knew personally. And a few years after his death, Ranke's influence started to shape the perspectives of German historical scholarship. With all these men Clausewitz shared certain broad assumptions that were

[63] See esp. R. L. F. Meinecke's celebrated works: *Cosmopolitanism and the National State* (Munich & Berlin, 1907; Princeton, 1970), at once an account of the process outlined above, and the most prominent explication of the German historical position by one of its greatest representatives; the same twofold significance belongs to his *Machiavellism, the Doctrine of raison d'etat and its Place in Modern History* (Munich, 1924; London, 1957), an expansion of the theme of the previous work, interpreting the political thought of the modern period, and written in a less optimistic mood after the First World War. Similarly illuminating and representative is E. Troeltsch, 'The Idea of Natural Law and Humanity in World Politics', App. to O. Gierke, *Natural Law and the Theory of Society 1500–1800* (Cambridge, 1934), 201–22. From the sea of literature written on the 'German Problem', see L. Krieger, *The German Idea of Freedom, History of a Political Tradition* (Boston, 1957); and G. G. Iggers, *The German Conception of History, the German Tradition of Historical Thought from Herder to the Present* (Middletown, Conn., 1968), a penetrating account and critique of the intellectual assumptions of the German historical school.

[64] Troeltsch, 'The Idea of Natural Law' in Gierke, *Natural Law*, p. 203.

to be common to the German conception of the state. In general terms these basic assumptions were: by and large, the state was the framework in which civilized communities developed; internally, the state was the higher and unifying expression of communal life; externally, owing to the natural dynamics in a society of sovereign entities, the interaction between states was governed by considerations of *raison d'état* or *Realpolitik*; within such a framework of relations, war had an integral part.

To do justice to Clausewitz's political and ethical outlook, which is mostly indirectly stated in his works, wider space than this volume can offer is obviously required. In the following discussion, therefore, only the principal themes of this outlook are mentioned in their relation to Clausewitz's view of war, and the first is war within the framework of political and international reality.

Clausewitz's concern with Machiavelli was part of a general revival of interest in Machiavelli in Germany, promoted most significantly by Fichte. This was one of the striking expressions of the new political attitudes that Clausewitz shared with the generation who witnessed the Napoleonic wars and the awakening of German national sentiment.[65] We have already noted some aspects of his affinity to Machiavelli, especially the importance placed on the vitality and dynamism of the political community, the call for great policies, and the rejection of half-measures. Clausewitz and Fichte, like most of their contemporaries, applied the darker side of Machiavelli's teaching to foreign affairs.[66] As noted by Paret, they both believed that Machiavelli's ideas on the relationship between the prince and his subjects mainly reflected his own political conditions of the Italian Renaissance, and no longer suited the enlightened societies of their own time. However, they thought that in the relations between states, where no law was in force, Machiavelli's conceptions were penetrating. 'Those who affect disgust for his [Machiavelli's] principles', wrote Clausewitz in 1807–8, 'are a kind of humanistic "petit-maîtres". What he says about the princes' policies toward their subjects is certainly largely outdated, because the condition of states have very much changed since his times . . . But this author is

[65] See esp. A. Elkan, 'Die Entdeckung Machiavellis in Deutschland zu Beginn des 19 Jahrhunderts', *Historische Zeitschrift*, CXIX (1919), 427–58.

[66] For a comprehensive discussion of Clausewitz, Fichte, and Machiavelli see again: Paret, *Clausewitz*, pp. 169–79.

especially instructive in regard to foreign relations.'[67] These are governed by considerations of *raison d'état*, and are dominated by the direct and implied use of force. Enlightened people, Fichte maintained, must face this reality.[68]

This view, which combined fundamental attitudes with an evaluation of the international reality in Europe, also found consistent expression in Clausewitz's analyses of contemporary political questions. In 'Umtriebe' written in the early 1820s, Clausewitz criticized the Romantic and liberal demands for the national unification of Germany. In a remarkable anticipation of the events of 1848 and 1866–70, he wrote:

> Germany can reach political unity in *one* way only, through the sword, when one state subdues all others. The time has not arrived for such subjugation, and if it should ever come it is impossible to predict at present which of the German states will become master of the others.[69]

Two works written by Clausewitz in the last year of his life, 'The Conditions in Europe Since the Polish Partitions' and 'Reducing the Many Political Questions that Preoccupy Germany, to the Question of Our Existence', are classic examples of what Meinecke described with satisfaction as the growing recognition in Germany of the primacy of *raison d'état* in political reality, as opposed to the old conceptions of cosmopolitan liberalism. In these works, Clausewitz discussed the questions preoccupying educated public opinion in Germany during the revolutions of 1830–1, in particular the Polish, Belgian, and Italian demands for unification and national independence which had been received with sympathy. His historical and political analysis focused exclusively on the *Realpolitik* considerations of Prussia and Germany, with total disregard not only for humanitarian concerns but also for any political considerations other than those derived from *raison d'état*, such as the domestic and social implications with which the events of 1830–1 were obviously imbued. According to Clausewitz, it was necessary for the

[67] Rothfels (ed.), *Schriften*, p. 63. For a similar way of thinking in the case of Fichte, see Paret, *Clausewitz*, p. 175.

[68] It is only too instructive that Meinecke describes Machiavelli as the first to reveal the true nature of the relation between states (*The Doctrine of raison d'état*, esp. ch. 1), and points to his revival in Germany with Fichte as marking a sobering process in German political thought, from 18th-cent. conceptions to a correct view of the state's true role (*Cosmopolitanism*, esp. ch. 6).

[69] 'Umtriebe' in Rothfels (ed.), *Schriften*, p. 171.

German people to oppose the independence of Poland, Belgium, and Italy because of those countries' natural allegiance to France. Their independence would severely impair the traditional German interest in preventing French hegemony in Europe.[70]

That these works were clearly directed against the prevailing views of liberal public opinion, was revealed by Clausewitz himself:

> I sought to make it clear to the good people that something besides *cosmopolitanism* should determine our position on the Belgian, Polish and other questions, that German independence was in the gravest danger, and that it was time to think about ourselves.[71]

This approach, which Clausewitz regarded as exposing reality as it actually was, as against liberal illusions, in fact incorporated a strong political preference for, and to a large extent was itself an unconscious expression of, a particular ideological point of view peculiar to the new attitudes in Germany.[72] The analysis in terms of *raison d'état* indeed called attention to certain characteristics of international relations, but also both explicitly and implicitly advocated political aims that focused on the power and stability of the state. This was the essence of Clausewitz's political outlook both on domestic affairs, which will be discussed later, and on foreign relations.

The new attitudes towards the essence and role of the state found their classic philosophical formulation in Hegel's *The Philosophy of Right*, one of the most influential works of political philosophy in the nineteenth century, published in Berlin in 1821. The affinity of Clausewitz's ideas to this famous work has, unfortunately, fallen victim to the general confusion concerning Hegel's influence on Clausewitz. Clausewitz was not a Hegelian, but some of the opinions which he had held from his youth and which had dominated his intellectual milieu appear to have received a definitive and distinctive conceptualization under the influence of Hegel's ideas.

According to Hegel, social ethics are the result of the general and unifying point of view achieved within the framework of the state.

[70] 'Die Verhältnisse Europas seit der Teilung Polens' and 'Zurückführung der vielen politischen Fragen, welche Deutschland beschäftigen, auf die unserer Gesamtexistenz' in Rothfels (ed.), *Schriften*, pp. 222–38. For a full discussion see Paret, *Clausewitz*, pp. 406–9.

[71] 21 Feb. 1831, in Schwartz, *Leben*, ii. 313; cited by Paret, *Clausewitz*, p. 406; my emphasis.

[72] This point was forcefully made in relation to the whole of the German tradition by Iggers, *The German Conception of History*, esp. p. 17.

Since in international society there exists no supreme authority which could enforce such norms of behaviour, so-called international law can only be a pale copy of the intra-state system, and is dependent on the good will of the states involved.[73] 'Attached to force', wrote Clausewitz at the opening of *On War*, 'are certain self-imposed, imperceptible limitations hardly worth mentioning, known as international law and custom, but they scarcely weaken it . . . moral force has no existence save as expressed in the state and the law.'[74]

From this view of the international system and the limits of moral order stems also the role and moral status of war. As mentioned above, the belief that this question finds no expression in Clausewitz's work, does not take into account the full scope and context of his world-view. Since in the international arena the rule of law does not exist and the prevailing behaviour is of almost unrestrained individualism, war is inherent in the system. 'People in our contemporary states must naturally love peace and hate war,' wrote Lossau; but states have interests that generate conflicts, 'and since no tribunal can resolve their conflicts, they seek justice by themselves. *Wars are therefore the exterior means of states to achieve by violence what they cannot achieve by peaceful means.*'[75] So long as this is the case, judging war by moral standards of measurement, derived from the intra-state reality of the civilized nations, would be pointless and wishful thinking and cannot be harmonized with reality. It was characteristic of the German Movement to reject any unsubstantiated attempt to shut out major parts of reality with kind-hearted ideals and standards of measurement of universal pretension.[76] In his *Apologie de la guerre* (1813), directed against Kant and the ideas of the eighteenth century, Rühle von Lilienstern, influenced by his friend Adam Müller, justified war by the realities of human behaviour and political life.[77] Furthermore, the idea that war also had a positive role to play in the development of civilization, and that it might even

[73] G. W. F. Hegel, *The Philosophy of Right* (Oxford, 1942), 'International Law', articles 330–40, pp. 212–16.

[74] *On War*, I, i, s.2, p. 75.

[75] Lossau, *Der Krieg*, p. 3; for the famous dictum see also p. 4.

[76] 'War', wrote Hegel, 'is not to be regarded as an absolute evil and as a purely external accident, which itself therefore has some accidental cause, be it injustices, the passions of nations or the holders of power, etc., or in short, something or other which ought not to be. It is to what is by nature accidental that accidents happen.' Thus, 'wars occur when the necessity of the case requires'. Hegel, *The Philosophy of Right*, article 324, p. 209; addition, p. 296.

[77] See above, Ch. 7. I, n. 44.

have an essential role in strengthening the social body, was also characteristic of the German Movement.[78] This too implied that the evaluation of war according to the categories of accepted social morals was narrow-sighted and inadequate.

This widely held view found a striking expression even with a prominent humanist such as Wilhelm von Humboldt, one of the chief reformers and a personal acquaintance of Clausewitz. In his *Limits of the State* (1791), Humboldt writes that war is 'one of the most wholesome manifestations that plays a role in the education of the human race'. War 'alone gives to the total structure the strength and the diversity without which facility would be weakness and unity would be void'. Anticipating the reformers, he wrote that professional armies should be replaced by a national army in order to 'inspire the citizen with a spirit of true war'.[79] In a memorandum concerning the army's budget in 1817, Humboldt, then Prussia's minister of education, listed the 'influence on the character of the nation' among the contributions of a strong army.[80]

'In times of peace', wrote Hegel, 'the particular spheres and functions pursue the path of satisfying their particular aims and minding their own business . . . In a situation of exigency, however, whether in home or foreign affairs, the organism of which these particular spheres are members fuses into [one].' Thus, 'war is the state of affairs which deals in earnest with the vanity of temporal goods and concerns . . . Corruption in nations would be the product of prolonged, let alone "perpetual", peace.'[81] Clausewitz expressed this *Zeitgeist* in almost identical terms. In *On War* he wrote the following passages, which are hardly ever cited:

Today practically no means other than war will educate a people in this spirit of boldness . . . Nothing else will counteract the softness and the desire

[78] See even Kant, 'Idea for a Universal History from a Cosmopolitan Point of View', and 'Perpetual Peace', in *Kant On History* (Indianapolis, 1963), 15–16, 19, 110–11; cited by Iggers, *The German Conception of History*, p. 47.

[79] Wilhelm von Humboldt, 'Ideen zu einem Versuch die Grenzen der Wirksamkeit des Staats zu bestimmen' in *Gesammelte Schriften* (Berlin, 1903–36), i. 137; cited by Iggers, *The German Conception of History*, p. 97. The book was published in full only in 1851, but parts of it, including the ch. on war appeared immediately after being written; Iggers, p. 297 n. 30. Cf. Humboldt with Clausewitz's letter to Fichte, p. 184 of this work.

[80] Humboldt, *Gesammelte Schriften*, xii. 170; cited by Iggers, *The German Conception of History*, p. 54.

[81] Hegel, *The Philosophy of Right*, article 278, p. 180–1; article 324, p. 210, and addition, p. 295.

for ease which debase the people in times of growing prosperity and increasing trade. A people and nation can hope for a strong position in the world only if national character and familiarity with war fortify each other by continual interaction.[82]

In his attitude towards the place of war within the human reality, Clausewitz was also, therefore, a true child of his times, and reflected the transformation of German national and political consciousness at the turn of the nineteenth century. Since his youth he had firmly believed—though he did not formulate this systematically—that the international arena was dominated by the behaviour of sovereign states guided by considerations of *raison d'état* in which power played the major role. In this reality, war was an immanent phenomenon, and perhaps also one which was not lacking in advantages; judging it by ethical categories taken from the social context was therefore pointless.

This outlook, which regards the state as the central organ of political reality, and which reflects the patriotic ideal that guided Clausewitz throughout his life, is also manifest in his views on the internal politics and structure of the state.

As mentioned earlier, the subordination of the military command to the political leadership was regarded by Clausewitz as a direct implication of the close link he discerned between political aims and military operations. In what was in fact largely a reaction against his own previous positions, he argued that it was erroneous to assume that once war was declared, the political leadership had to give the army command a free hand and all available means for purely military planning. There was no such thing as purely military planning; military planning was derived from the political aims of the war.[83] The relationship between political aims and military means was, of course, not one-sided. The means had to suit the ends, but the ends too could not be divorced from the available means; 'the political aim is [not] a tyrant', and the politician should not 'issue orders that defeat the purpose they are meant to serve'.[84] A continuous interplay exists between the aims and the means.[85]

[82] *On War*, III, 6, p. 192.
[83] Ibid. VIII, 6B, p. 607.
[84] Ibid. I, 1, s. 23, p. 87; VIII, 6B, p. 608. This point was also stressed in a famous letter which Clausewitz wrote in 1827 to the then chief of staff General Müffling, and which was to be cited by Moltke during his clash with Bismarck; for the letter see *Two Letters on Strategy*, ed. and trans. by P. Paret and D. Morgan (Carlisle, 1984).
[85] For an elaboration of this relationship, see *On War*, VIII, 6B, pp. 607–8; ibid. I, 1, s. 23–4, p. 87.

Yet, this discussion also provides us with a further insight into Clausewitz's political outlook and conception of the state. The summary of the relationship between political leadership and military command in purely instrumental terms is characteristic. 'No conflict', he wrote, 'need arise any longer between political and military interests—not from the nature of the case at any rate—and should it arise it will show no more than a lack of understanding.'[86] The fact that political and military establishments consist of people who in real life may differ and even clash, not only over the matching of aims and means but also over the desired political values, objectives, and directions of action themselves, seemingly does not occur to Clausewitz.

The ideal that guided Clausewitz throughout his life—the vitality, stability, and power of the community in its political framework—was a characteristic product, historically and ideologically, of the continuous rise of the centralized state throughout the early modern period and its triumph over all other social focuses of power. As a result of this, all independent armed forces were also incorporated into a central army with a purely instrumental role. This historical development and corresponding ethos had particular significance in Prussia, where, since the time of the Great Elector, they had been responsible for the transformation of the Hohenzollern state from a poor principality into a major European power. Both the development and ethos reached their peak with the perfection of absolutism under Frederick the Great; and were transformed by the resistance to French political ideas and occupation, reappearing in an updated and more comprehensive form, with an emphasis on the corporative nature of the nation and state.

Clausewitz's political position when the French Revolution shattered the *ancien régime*, and when the question of social and political constitutions was at the centre of the European agenda, must be borne in mind. What in fact was his exact ideological attitude to the major social and political currents of his period? This point remained somewhat unclear, and was thus addressed by Behrens in 'Which Side was Clausewitz On?'.[87] For themselves, Clausewitz did not share the aspirations of the Third Estate nor the defensive

[86] Ibid. VIII, 6B, p. 607.

[87] Behrens's principal point was largely accepted by Paret, who expanded it in 'Die politischen Ansichten von Clausewitz', in Clausewitz Gesellschaft (ed.), *Freiheit ohne Krieg*, pp. 332–48.

position of the old ruling classes; he held neither a republican, constitutional, nor absolutist position. He was guided by a passionate political vision derived from the traditional Prussian political ethos and reinforced by the new organic and evolutionary view of society and the nation: this political vision aimed at the greatness and well-being of the people and state as defined in terms of stability, vitality, and power.

Like Scharnhorst he was one of the exponents of the Prussian reform movement because he thought that the *ancien régime* no longer suited the conditions and needs of the period, and believed that the expansion of the social basis of the Prussian state was essential for her continued survival, independence, and status of power in the post-Revolutionary era.[88] For these reasons he relentlessly defended the *Landwehr* during the political struggles of the Restoration which culminated in W. von Humboldt's resignation in 1819, and even at one time proposed a form of parliamentary institution.[89] Yet, only a short while later and for the very same reasons, he came out strongly against the nationalist and liberal unrest and even opposed the demands for a constitution and parliament, which he regarded in the early 1820s as divorced from Prussia's present reality and dangerous for her stability and well-being.[90]

Parliament was for Clausewitz predominantly a means for social cohesion through the expansion of the government's base of support. Similarly, in his arguments for the *Landwehr*, the main target for the attacks of the forces of reaction, he attempted to show that its necessity for the defence and international position of Prussia far outweighed, and even made irrelevant, any particular social or class argument for, or especially against, it. This was not merely a clever way to evade what was actually the core of the problem, as Paret appears to imply.[91] Nor was the criterion that Clausewitz used an

[88] See Krieger, *The German Idea of Freedom*, pp. 196–202, which is better on Gneisenau than on Clausewitz.

[89] 'Über die politischen Vortheile und Nachteile der Preussischen Landwehr', written at the end of 1819; in Schwartz, *Leben*, ii. 288–93; for the idea of parliament, to expand the government base of support, see ibid. 291. The argument for the *Landwehr* is further elaborated on in 'Unsere Kriegsverfassung' written at the same period; Rothfels (ed.), *Schriften*, pp. 142–53.

[90] See 'Umtriebe' in Rothfels, (ed.) *Schriften*, pp. 153–95, called by Paret—in view of what appears as a change of political positions by Clausewitz—'the most puzzling of all of Clausewitz's works'; Paret, *Clausewitz*, p. 299.

[91] Ibid. 295.

objective standard of measurement. Rather this approach reflected his particular political attitudes which subordinated any social and political ideal or objective to what he regarded as the true interest of the state and people. As pointed out by Behrens and followed up by Paret, Clausewitz basically maintained the eighteenth-century, Frederickian, paternalist view of politics. And this was reinforced by the new corporate conception of society. As he put forward in 'Umtriebe', the welfare of the people had to be the government's main concern; furthermore, the new conditions required a closer involvement of the people in government, than had been the case in Frederick's time, as a means of social cohesion. However, the actual conduct of politics was not a matter for particular interests but was to be firmly held in the hands of the government which had to be guided by what Clausewitz regarded as the general great interests of society as a whole.[92]

In the very years in which Clausewitz wrote 'Umtriebe', Hegel gave the prevailing view of the state its supreme philosophical expression in *The Philosophy of Right* (1821). According to Hegel, the various groups and interests contending in civil society, the sphere of the war of all against all, find their ethical and rational solution in the state. The leadership of the state remains above the struggle of the particular forces. It embodies unity, disinterest, the supreme expression of society as a whole. The similarity of this conception to Clausewitz's trend of thought is obvious. Unfortunately we only know that 'Umtriebe' was written sometime in the early 1820s, but whether before or after 1821 is unclear. Consequently it is very difficult to determine whether Clausewitz's own ideas were indeed reinforced and influenced by Hegel's celebrated conceptions or simply expressed, independently, very similar intellectual trends, common to the German movement.

Be that as it may, Clausewitz again alluded to his own political outlook in Books VIII and I of *On War*, written in the late 1820s. As we have seen, at that time it is highly improbable that he was not familiar with *The Philosophy of Right*, and indeed there appear to be some distinct features that suggest its influence. In explaining the supremacy of the political over the military, Clausewitz wrote:

> It can be taken as agreed that the aim of policy is to unify and reconcile all aspects of internal administration as well as of spiritual values, and

[92] In 'Umtriebe' see esp. pp. 176–7. Also see Behrens, '*Which Side*' and Paret, 'Die politischen Ansichten von Clausewitz', pp. 340–2.

whatever else the *moral philosopher may care to add*. Policy, of course, is nothing in itself; it is simply the trustee for all these interests against the outside world. That it can err, subserve the ambitions, private interests, and vanity of those in power, is neither here nor there . . . here we can only treat policy as representative of all interests of the community.[93]

Later, in describing the 'remarkable trinity', Clausewitz actually refers to policy as 'reason' governing the passions of the people and the activity of the army.[94]

In the last stages of the writing of *On War*, Clausewitz's attitudes to politics and the state were therefore more formally conceptualized. These attitudes reflected deep-rooted traditions embedded both in the Prussian historical context and in the distinctive character of the German political philosophy. They were very probably influenced in the last decade of Clausewitz's life by Hegel's highly renowned ideas. Controversies over state policy were regarded by Clausewitz as a problem of interpreting a rational common political interest (providing, of course, they did not 'subserve ambitions, private interests, and vanity'), rather than as a struggle between contending political visions and objectives within the state. As Gerhard Ritter points out, Clausewitz did not acknowledge the possibility of an existential gap between different aims in society.[95] The rejection of the atomistic view of society for an organic and rational harmony of interests was central to the ideas of the German movement. In this context the relationship between political leadership and military command was also understood in purely instrumental terms.

Yet, Clausewitz's own life story not only sets this conception into its historical context but also places it in an ironic light. Throughout the great events of his period, the struggle for independence against Napoleon and the reform of the Prussian state, Clausewitz, the military man, bitterly opposed the political aims and even the declared policy of his king. He and his fellow reformers in the army, who comprised a 'purely military body', took part not merely in discussions on the adjustment of aims and means, but in a formidable

[93] *On War*, VIII, 6B, pp. 606–7; my italics. These ideas undoubtedly formalized earlier notions. Compare with Lossau, *Der Krieg*, p. 7: 'Politics operates for the existence and external prosperity of states. It safeguards the individual interests equally; it determines the fundamental idea, the direction, and the aim that the state should advance.'

[94] *On War*, I, 1, s .28, p. 89.

[95] Gerhard Ritter, *The Sword and The Scepter* (Miami, 1969, IV), i. 67.

power struggle within the Prussian leadership, which stemmed from conflicting class interests and contending social and political visions, and which centred on no less than the reshaping of Prussia's social structure and foreign policy. At a time of crisis, Clausewitz left for Russia to fight Napoleon, acting against government policy and his king's orders. The fundamental controversies in reality encompassed a much wider scope than could be resolved by *raisons d'état*, and cut across the institutional lines of political leadership and military command.

This irony has escaped those who today have raised to prominence Clausewitz's conception of the relationship between political leadership and military command. They have had in mind the controversies between Bismarck and Moltke, and Truman and MacArthur, where a rejection of the particular positions held by the military command was happily in union with our contemporary political outlook that postulates the supremacy of the political leadership.

Indeed, the significance of Clausewitz's conceptions of the relationship between political leadership and military command largely derives from the role these conceptions have played in supporting the political outlook of today.

During the Franco-Prussian War, in his famous clash with Bismarck, Moltke formulated the general staff's claim to a shared authority with the *Kanzler* in the leadership of the state, under the king's supreme authority. In this he gave expression to the relationship between political and military leadership that was embedded in the political structure and ethos of the Second Reich. In the 1930s Ludendorff expressed the natural point of view of a militarist value-system when he declared Clausewitz obsolete, and made the political leadership an instrument of the military command for the harnessing of civilian life to the needs of war.[96]

National self-examination after the Second World War led the German historians to Clausewitz, whose conception of the relationship between political leadership and military command could be integrated into the new liberal-democratic ideal. This conception, divorced from its actual historical and intellectual context, and sharply contrasted with the legacy of the Second Reich, became

[96] E. Ludendorff, *The Total War* (London, n.d.).

one of the major reasons for Clausewitz's revival. In the United States, the complex problems of controlling the military machinery in a superpower democracy led to a similar trend.[97]

A certain compatibility in viewing the relationship between political leadership and military command was thus responsible for the fact that the conceptions of a Prussian thinker, whose political thought centred on adapting the tradition of Prussian *étatisme* to the conditions of nationalist post-Revolutionary Europe, were enlisted to serve the political and ideological code of the liberal Western democracies.

[97] The most notable example for the renaissance of Clausewitz's ideas in this context is Bernard Brodie's *War and Politics* (New York, 1973).

CONCLUSION

This book really requires two conclusions, to match the two major arguments raised in it. The first is concerned strictly with the interpretation of the core of Clausewitz's military thought, his conception of the nature of war. The second deals with the wider intellectual framework of military thought, as presented in relation to the two periods described, and the implications of this presentation for the understanding of military theory in general. About this second topic in particular, there is much more to be said than I can possibly hope to discuss here. A few words of conclusion will have to suffice.

From the outset, there was a latent tension in Clausewitz's thought between his historicist sense and particularist notions on the one hand, and his universalist quest on the other. This tension surfaced in 1827, calling into question some of Clausewitz's ideas regarding defence and attack, and rapidly expanding to threaten his conception of the nature of war. Henceforth, his thinking underwent a process of continuous transformation which was terminated only by his death. Had it been carried further, this process had the *potential* to demolish most of the surviving components of Clausewitz's lifelong conception of war. Indeed, this is why Delbrück was able to rely on Clausewitz's ideas in rehabilitating eighteenth-century warfare, and why modern interpreters could often disregard Clausewitz's emphasis on the clash of forces in combat. In both cases, however, Clausewitz himself never went so far.

In the event, his intellectual development in his final years introduced a great deal of 'mystification', and it is very doubtful whether he would have retreated from this direction. This appears to be so because, apart from its success in incorporating his old ideas with his new ones, his later intellectual structure had the great appeal of satisfying his deep psychological need to give his work the form expected from a 'truly philosophical' treatment of war.

The tensions in Clausewitz's own work resulted in corresponding interpretative antinomies. Thus, for example, his work was regarded both as an analysis of contemporary Napoleonic warfare and, at the same time, as a universal theory of war. Closely linked is the preposterous idea that Clausewitz was concerned with the nature of war, as distinct and remote from any normative approach to the actual conduct of war. Nothing could be further removed from Clausewitz's own motives and work throughout his life. Another example, mentioned before, is to be found in Clausewitz's attitude to eighteenth-century warfare. Since this warfare has been rehabilitated in both historical and strategic thinking, and since Clausewitz was not to be accused of harbouring an unhistorical approach, his attitude had to be presented merely as a criticism of the *excesses* of the war of manœuvre.

The endemic difficulties in interpreting Clausewitz have stemmed largely from the fact that *On War* is a classic case where the text cannot be understood without its context; not only the military and intellectual context but also that provided by the evolution of Clausewitz's own thought. The opening part of *On War* reflects in effect the latest stage in his development, while the middle reflects the earliest, and the last the intermediate, each incorporating fundamentally contrasting ideas. In short, reading *On War* as it stands, without the necessary preliminary knowledge, is bound to result in misunderstanding.

Although aware of the unfinished state of the work, and to some degree cognizant of its internal development, many of Clausewitz's interpreters have still attempted to explain *On War* as a coherent whole. When coupled with our contemporary attitudes and sensitivities, this has often led to a harmonizing interpretation and partisan approach, with the real Clausewitz sterilized and almost disappearing behind mountains of scholarly talk.

Conversely, it is clear that the men of the nineteenth century (and for that matter also Liddell Hart) were not so ridiculously mistaken in their understanding of Clausewitz as it has become the fashion to believe. While being perhaps slightly more nationalist and militarist than him, they were organically—both historically and intellectually—far closer to him than the men of our era ever could be.

Much of Clausewitz's reputation as a profound thinker has therefore resulted from the confusion among his interpreters. In a sense, Clausewitz could never have been wrong or less than profound because no one could be quite sure that he understood the true meaning of Clausewitz's ideas. Yet, Clausewitz's real intellectual greatness and one reason for the living interest in his work stems from a unique achievement that has never been equalled. He offered a most sophisticated formulation of the theory of war, based on a highly stimulating intellectual paradigm, and brought the conception of military theory into line with the forefront of the general theoretical outlook of his time.

Delineating the subsequent career of the two intellectual traditions described in this book would require another volume. One tradition, in close affinity to the scientific enterprise, went through positivism to logical positivism and its descendants in the social sciences. The other tradition, stressing the gulf between the sciences and the humanities and the dominance of history and man's inner world over the latter, was similarly carried forward by the German Movement of the nineteenth century to our contemporary contentions. These underlying historical trends have received too little attention.

In this book in any case, the primary aim is not to strike a new balance between the two theoretical traditions that have dominated modern military thought, though in many respects such a balance may certainly be implied. Nor is it to bring the one into, or remove the other from, the scene of contemporary strategic thinking, though the striking resemblance of their arguments to the modern debate between the 'traditional' and 'scientific' schools in the social sciences makes their intellectual legacy appear remarkably relevant. While the ideas of the military thinkers of the Enlightenment in particular were poorly understood and caricatured, and while it has been stressed in this work that human thinking takes a variety of forms, a historical approach does not imply an equal acceptance of all ideas. It should, however, make the strange familiar and comprehensible.

A great deal of progress has been made in understanding war in its wider contexts, particularly the social one. Yet, this development has barely touched the intellectual sphere, with unfortunate consequences for the study of military thought. Hence the main point of this book, at once historical and theoretical; in it an attempt has been made to reject the 'naïve' approach to military thought.

Historically, Clausewitz's ideas did not appear out of thin air, nor were his predecessors curious eccentrics with peculiar ideas, as the German military school would have us believe. Historians have been largely unaware of the historical traditions that have predetermined their view of the period. Our story is in fact merely one aspect of an old and well-known story: the conflict between the Enlightenment and the German Movement. Both the works of the military thinkers of the Enlightenment and those of Clausewitz were strikingly comprehensive expressions of the general manner in which the intellectual élites of Western civilization in two successive periods understood and interpreted their world.

The theoretical point is closely related: that *what* people think cannot be separated from the question of *how* they think, or from the circumstances in which they operate and to which they react. Military theory is not a general body of knowledge to be discovered and elaborated, but is comprised of changing conceptual frameworks which are developed in response to varying challenges, and which always involve interpretation, reflecting particular human perspectives, attitudes, and emphases. Consequently there is no such thing as a 'theoretical', 'positivist' understanding of past military theories 'as they are', nor is there much sense in discussing them 'abstractly' or judging their value without keeping in mind the historical and intellectual circumstances in which they were formed. The theoretical premises of every conception of military theory cannot but depend on some overall (albeit unconscious) picture of the world.

APPENDIX

Clausewitz's Final Notes Revisited

Among his literary remains, Clausewitz left us two notes written at an advanced stage of his work on *On War* which describe the state of the treatise and his plans for its future development. These notes are highly important for the understanding of Clausewitz's intellectual career, particularly because of the comprehensive revision of his work which he undertook, but did not complete, in the last years of his life. Unfortunately, only one of these notes, albeit the most important one, announcing the planned revision, is dated: 10 July 1827. The other bears no date.

Clausewitz's wife, Marie, who, with the assistance of her brother and Major O'Etzel, published Clausewitz's posthumous works, made no attempt to date this note or to connect it to any specific event.[1] However, apparently she tended to believe—though was careful not to determine—that it was written subsequent to the note of July 1827, and placed it after this note at the opening of Clausewitz's *Collected Works*. She wrote that it 'appears to belong to a very late date'[2].

A century later, Clausewitz's interpreters were much bolder. Endorsing the prevailing view, they decided that the undated note must have been composed in 1830 when we know from Marie that Clausewitz, who had been transferred from his post as the Director of the War School in Berlin to field service in the artillery, had been obliged to stop his work on *On War* and had packed and sealed his papers until time allowed him to resume writing. This date has had

[1] e.g. she was much more prepared to commit herself in the case of a considerably older note. She attributed it (and apparently rightly so) to Clausewitz's period in Koblenz in 1816–18 when he wrote the early concise work which was to lead to the composition of *On War*; see *Hinterlassene Werke*, vol. i, p. viii.

[2] Ibid., p. xix.

much appeal not least because it created the somewhat romantic picture of a fateful moment when Clausewitz, upon leaving his work to which he was never to return, left a final record of his intentions. From the 1930s there has therefore been a consensus among Clausewitz's interpreters regarding the dating of this note.[3]

What I would like to argue here is that this dating is highly improbable, that it created some difficult problems that scholars have failed to resolve, and, even more importantly, that it is conspicuously divorced from all that we know about Clausewitz's later development. I will suggest that the undated note was written in fact shortly *prior* to the note of July 1827, possibly only a few months before.

In the note of July 1827, Clausewitz assesses the far-reaching implications on his work of his new discovery that there are two types of war, absolute and limited, and that war is the continuation of policy by other means. With most of *On War* already written, he now must 'regard the first six books which are already in a clean copy merely as a rather formless mass that must be thoroughly reworked once more'.[4] He therefore states his intention to work in the light of his new guidelines on Books VII and VIII, in both of which he has only sketched or outlined several chapters ('*entworfen*' '*Skizzen*' for Book VII, and '*entworfen*' for Book VIII). Then, after finishing the original plan of the work, he will return to revise the first six books.[5]

Now, the undated note was allegedly written in 1830. Yet it reveals no progress on the state of affairs described in July 1827. In fact, if it were not for the basic assumption, I would suggest that one would have had to admit that the undated note represents a slight *regression* on the note of July 1827. Let us examine the texts. In the note of July 1827, Book VI is undistinguished among the first six completed, though unrevised books of *On War*. By contrast, in the undated note, Clausewitz while appearing to single out Book VI

[3] The date and circumstances were suggested in the first rigorous treatment of the issue in Herbert Rosinski's 'Die Entwicklung von Clausewitz Werk "Vom Kriege" im Lichte seiner "Vorreden" und "Nachrichten" ', *Historische Zeitschrift* CLI (1935), 278–93. Despite other differences of opinion, this was accepted by Eberhard Kessel in 'Zur Entstehungsgeschichte von Clausewitz Werk "Vom Kriege" ', *Historische Zeitshcrift*, CLII (1935), 97–100, and reiterated by all of Clausewitz's later interpreters.

[4] 'Note of 10 July 1827', *On War*, p. 69.

[5] Ibid., pp. 69–70.

for especially harsh treatment, describes it as a mere attempt or sketch (*blosser Versuch*).[6] This is particularly puzzling since Book VI, as we know it, comprises more than a fourth of the whole work, and is 2.5 to 3.5 times as large as any of the other books of *On War*. Regarding Books VII and VIII, whereas in the note of July 1827 Clausewitz states that in both he has sketched or drafted several chapters, in the undated note he writes similar things only about Book VII (*die Gegenstände flüchtig hingeworfen*). About Book VIII he speaks only in the future tense, presenting its planned subject and nature. There is no sign in the text that any of this Book actually exists.[7]

Indeed, Clausewitz's modern interpreters have found the final notes somewhat problematic. Since they have assumed that the undated note was written in 1830, they have all agreed that for some reason or another, after three years, Clausewitz seems to have made very little progress. Furthermore, it appeared that Clausewitz had almost completely disregarded his working plan of July 1827. He did not, or barely (there are slightly different opinions here) work on Books VII and VIII which are, at best, described in both notes in fairly similar terms. Instead, we are told, he went directly to the revision of his first books, where he did some work on Book I and perhaps also on a few others. In his work on *On War*, this is all that he achieved between 1827 and 1830.

Let us begin with Books VII and VIII as we know them. Marie believed that they were indeed unfinished and in a state of rough sketches.[8] We have an important clue as to why. From Clausewitz's note of July 1827 we know that he produced clean copies of the first six books of *On War* which he previously regarded as more or less complete, possibly after finishing each of them. However, once he started the process of clarifying his new ideas, there was no point in doing that. Whatever work he did on Books VII and VIII, he apparently did not produce clean copies of them. Thus for Marie, the state of the manuscript, the evidence of the undated note, and her assumption about the note's 'very late' composition must have reinforced each other.

Now, Books VII and VIII were not copied into a clean version, but, in the state in which we know them, should they be described as

[6] Ibid., p. 70. [7] Ibid., pp. 69, 70.
[8] *Hinterlassene Werke*, vol. iii, p. v.

sketched in outline form? All of Clausewitz's interpreters have been obliged by their dating of the undated note to reply in the affirmative. But there are no real grounds for such a statement. If size is considered, then Book VII is very average, while Book VIII is the third largest in *On War*.[9] Content-wise, Book VII 'The Attack' appears quite complete, particularly as Clausewitz specifically states that it is only a supplement to Book VI 'Defence'.[10] Chapter 16 even deals with limited objectives and falls in step with his new guidelines. As for Book VIII, this is where Clausewitz elaborates his new ideas most fully and extensively.

Why then would Clausewitz in 1830 particularly describe the last three books of *On War* as sketches (if, indeed, he says even that about Book VIII)? Moreover, if after July 1827 he did not work on Books VII and VIII, how is it that they express his new ideas? Alternatively, if he did work on these books, why does the undated note reflect absolutely no progress on the note of July 1827? These contradictions have remained unresolved.

There is one crucial reason why all of Clausewitz's interpreters have believed that the undated note was written in 1830 and recorded the progress of his work-plan put forward three years earlier. In the undated note, Clausewitz wrote: 'The first chapter of Book One alone I regard as finished. It will at least serve the whole by indicating the direction I meant to follow everywhere.'[11] Since we know that the beginning of Book I indeed represents the latest stage in the development of Clausewitz's ideas, this statement has been perceived as the latest account of the revision of his work.

This brings us to the core of the argument. Interpreters who have had the note of July 1827 in mind, have overlooked something very fundamental. In the undated note, Clausewitz does not mention any revision nor does he even allude to the ideas of policy and war, or absolute, limited, and real wars. In short, he does not refer to what was in 1830 the focus of his work and his major concern for the future, to what is universally supposed to have been the whole purpose of the note. Indeed, assuming this purpose, Clausewitz's account appears strangely obscure. He fails to enlighten us about the things that were the most important to him. Instead he presents

[9] Book VII is larger than Books I, II, and III, roughly as large as Book IV, and smaller than Books VI, V, and VIII. Book VIII is only smaller than Books VI and V.

[10] *On War* VII, 1, p. 523.

[11] Ibid., p. 70.

a very long list of propositions which are intended to prove the possibility of a general theory of war, and which summarize major themes from *On War*. Curiously enough, the ideas of policy and war, and absolute, limited, or real war do not appear here either. The undated note could not have been written in 1830.

If this is so then how are we to understand Clausewitz's reference to the first chapter of Book I as the only one he regarded as finished? I would suggest that a remarkable coincidence was responsible here for the misinterpretation, but this is better explained from the beginning.

The undated note appears to have been composed shortly before the note of July 1827, possibly early in the same year. It was written when, and because, Clausewitz discovered that there was a problem in regarding all-out war as the only type of war and the sole foundation of theory. We know this happened while he was writing Book VI, or perhaps when he was already copying it. The whole of his intellectual enterprise now appeared in jeopardy.

Hence a dominant characteristic of the note, its melancholic tone. Apart from the fact that Clausewitz declares that most of his work is unsatisfactory and should be regarded merely as 'working material', he devotes the larger part of the note to the assessment of the question whether a theory of war, despite its 'extraordinary difficulties', is possible at all. Although his tone appears to be unusually subdued, he answers in the affirmative, relying on the list of propositions which he regards as universal and which are taken from the first books of *On War*.

This brings us to another dominant characteristic of the note: in Clausewitz's intellectual development it is patently archaic.[12] As mentioned above, there is no trace of his new ideas. The exciting and fundamental arguments of Book VIII cannot be found in the list of propositions. Interestingly enough, the opening theme in the list is the relationship between defence and attack, the subject of Book VI. Indeed, let us return to Clausewitz's account of his work in the note and examine it in the light of our new date.

Clausewitz describes Book VI as a mere attempt or sketch (*blosser Versuch*) and states that he will rewrite it entirely and look for

[12] This can already be seen in the clearly archaic title which Clausewitz uses for his manuscript: 'on the conduct of great wars' [*über die Führung des grossen Krieges*].

another 'way out' (*Ausweg*). Assuming as he did that the note was written in 1830, Aron believed that Clausewitz deemed it necessary to revise his views on defence and attack in the light of his new ideas on policy and war.[13] While this might be true, it still does not explain why Book VI is singled out. Surely the same revision was needed for all of Clausewitz's early books, particularly Book III dealing with strategy. However, once we redate the note, all this becomes much clearer.

When he was writing Book VI, Clausewitz encountered a major problem—the possibility of a defence with a limited aim. Not only did he now become dissatisfied with Book VI but he sensed that the problem might have a bearing on all his previous work. While there were now question marks on all his work, Clausewitz had not yet clarified to himself the exact nature and full implications of the problem, and still believed that at least the fundamentals of his work remained unaffected whatever the adaptations and additions he would have to make. In case of an early death, he wanted the world to know both sides of the coin. He therefore stated that the first chapter of Book I—which, as one would expect, in the early phase of *On War* also dealt with the question 'What is War?' and encapsulated Clausewitz's fundamental view on the subject—was the only one that he regarded as finished, and that it indicated the direction he wanted to follow everywhere.[14]

In the last two books of *On War*, only the main topics of Book VII were roughly sketched. In Book VIII there was the main idea but apparently nothing substantial written.[15]

Now, whereas Clausewitz wrote the undated note when he sensed that he had encountered a difficult problem, he composed the note of July 1827 when he began to clarify to himself the nature and implications of this problem and work out a solution for it. This proximity of time and subject is responsible for the remarkable similarity between the two notes which in many respects are almost a mirror image of one another. The sequence of events may have been as follows. Firstly, Clausewitz devised the idea of the two types of war, absolute and limited. Putting this idea into practice, he wrote the last three chapters of Book VI, thus finding—as he had stated he would do—another way out (*Ausweg*). He could then finish

[13] Aron, *Clausewitz*; pp. 239–50.
[14] 'Undated Note', *On War*, p. 70.
[15] Ibid.

copying the book. His new idea led him to develop the idea of the relationship between policy and war. In the note of July 1827 he tells us what he did next. He foresaw that 'the main application of this point will not be made until Book Eight'. He therefore 'drafted several chapters' of this new book, 'done with the idea that the labour itself would show what the real problems were'. Indeed, 'that in fact is what happened', and having clarified his mind to a large extent and having decided what needed to be done, he then wrote the note of July 1827.[16]

In this note he presented his two new ideas and their implications on his work. The first six books which were already in a clean copy would now have to be revised. However, he would first work on and finish the last two books. Book VII 'On Attack', which apparently remained in the state it had been when the earlier note had been written, would be revised and completed. After finishing this book, he would 'go at once and work out Book VIII in full'. This book, 'War Plans', is where his new ideas would really be elaborated.[17] This process would be of crucial importance:

> If the working out of Book Eight results in clearing my own mind and in really establishing the main features of war, it will be all the easier for me to apply the same criteria to the first six books . . . *Only when I have reached that point therefore*, shall I take the revision of the first six books in hand.[18]

How did this work-plan materialize between 1827 and 1830 when Clausewitz had to stop his work? As mentioned above, when the undated note was given the date 1830, all had to agree that Clausewitz appeared neither to have made substantial progress nor to follow his planned programme. Now we have no 'Final Note', yet the course of Clausewitz's work during his last three creative years becomes quite clear and consistent with the *On War* that we know. As he had stated he would do, he first worked on and completed Book VII 'On Attack', inserting his new idea of the two types of war in chapters 15 and 16. He then went on to write and finish the large Book VIII 'War Plans', the natural place and the real testing ground of the new ideas. Only after developing these ideas extensively and clarifying his thoughts in writing this book, did he undertake the revision of the first six books.

[16] Ibid., p. 69. [17] Ibid., pp. 69–70.
[18] Ibid., p. 70; my italics.

How much progress did he make in the revision of these first books? Schering, famous for his unsubstantiated speculations and tendentious fantasies, 'discovered', for example, in his latest work *Clausewitz, Geist und Tat* (1941) a new intellectual transformation centring on Book II. Rothfels believed that Clausewitz revised 'parts at least of Book I (probably chapters 1–3) and of Book II (certainly chapter 2)'.[19] Most interpreters hold more or less the same opinion. Yet, there is absolutely no evidence for this expansive hypothesis. There is nothing new in chapter 2 of Book II, and certainly not a trace of the ideas of the different types of war and policy and war. Moreover, their opinion is clearly contradicted by one of the solid pieces of evidence that we do possess and that has also been curiously overlooked.

In her introduction, Marie tells us specifically that in preparing Clausewitz's literary works for publication, her brother 'found the beginnings of the revision that my beloved husband mentions as a future project in the note of 1827 . . . The revisions have been inserted in those places [*Stellen*] of Book I for which they were intended (*they did not go further*).'[20] When we remember that the first books of *On War* were in clean copy, Marie's testimony becomes even clearer. In going back to revise these books Clausewitz apparently did not literally rewrite them completely. He merely rewrote, amended, and added sections (some of which were obviously quite extensive) to be incorporated into the existing text.

All this fits in perfectly with another piece of evidence. We know from Marie that in November 1831, after concluding his mission on the Polish frontier and shortly before his death, Clausewitz hoped to finish his work during the course of that winter.[21] If it is assumed, by dating the undated note 1830, that between 1827 and 1830 he made very little progress in writing Books VII and VIII and that these books remained but sketches, this hope appears peculiarly optimistic. However, once that basic assumption is abandoned, Clausewitz's hope is revealed in a new light. He wrote the now-extensive Books VII and VIII, and revised Book I which he had naturally anticipated—as we know from the note of 1827—to be the most affected by the revision and the application of his new ideas.[22] In the winter of 1831 he therefore believed that he was

[19] Hans Rothfels, 'Clausewitz', in Earle (ed.), *Makers of Modern Strategy*, p. 108.
[20] *On War*, p. 67; my italics. [21] Ibid., p. 66. [22] Ibid., p. 69.

mainly left with the incorporation of these ideas into the rest of the text of *On War*—surely no small task, but far smaller than the one interpreters have assumed.

If indeed we no longer possess a 'Final Note' with Clausewitz's own testimony, then which parts of *On War* did he regard as truly finished before his death? This question appears to me to be somewhat misleading, because after the intellectual transformation of July 1827, Clausewitz's ideas were undergoing continuous development. The idea of limited war which appeared at the end of Book VI and which was later incorporated into Book VII, was supplemented in the note of July 1827 by the idea of the relationship between policy and war. Both ideas were then worked out in Book VIII where Clausewitz continued to elaborate his thoughts. Chapters 1 and 2 of Book I reveal a further development where Clausewitz no longer regards absolute war as superior to real war. There is no evidence that the revision went any further in Book I and there was indeed no reason why any of the other chapters of this book should have been affected by Clausewitz's new ideas. As it is, Books II–V of *On War* were completely unrevised but the revision was probably needed most badly in Book III on strategy. In the more advanced part of *On War*, it is doubtful whether Clausewitz deemed further considerable revision of Books VI and VII necessary, but he probably wanted to bring Book VIII into line with the latest developments of Book I. Whether Clausewitz would have developed his new ideas further and in new directions, thus generating new changes throughout his work, no one can tell.

SELECT BIBLIOGRAPHY OF WORKS CITED

Works Composed before 1837

BERENHORST, GEORG HEINRICH VON, *Betrachtungen über die Kriegskunst, über ihre Fortschritte, ihre Widersprüche und ihre Zuverlässigkeit (3rd edn.; Leipzig, 1827), including Randglossen* and *Aphorismen.*

—— *Aus dem Nachlasse*, ed. E. von Bülow (2 vols.; Dessau, 1845, 1847).

BINZER, J. L. J., Über die militärischen Werke des Herrn von Bülow (Kiel, 1803).

BRENCKENHOFF, L. S. VON, *Paradoxa, gröstentheils militärischen Inhalts* (n.p., 1783).

BÜLOW, A. H . D. VON, *The Spirit of the Modern System of War* (London, 1806).

—— *Der Feldzug von 1800, miliärisch-politisch betrachtet* (Berlin, 1801).

—— *Neue Taktik der Neuern, wie Sie seyn sollte* (Leipzig, 1805).

—— *Lehrsätze des neuern Krieges, oder reine und angewandte Strategie aus dem Geist des neuern Kriegssystems* (Berlin, 1805).

—— *Der Feldzug von 1805, militärisch-politisch betrachtet*, (n.p., 1806).

—— *Militärische und vermischte Schriften*, ed. E. Bülow and W. Rüstow (Leipzig, 1853).

—— *Pacatus Orbis* (London, 1867).

CARL VON OESTERREICH [Archduke Charles], *Ausgewählte Schriften*, ed. F. X. Malcher. (6 vols.; Vienna and Leipzig, 1893–4).

CLAUSEWITZ, CARL VON, *Hinterlassene Werke* (10 vols.; Berlin, 1832–7).

—— *On War*, ed. and trans. M. Howard and P. Paret, introductory essays by P. Paret, M. Howard, and B. Brodie, with a commentary by B. Brodie (Princeton, 1976).

—— *Vom Kriege*, ed. W. Hahlweg (16th edn.; Bonn, 1952).

—— *Verstreute klein Schriften*, ed. W. Hahlweg (Osnabrück, 1979).

—— *Carl von Clausewitz, Politische Schriften und Briefe*, ed. H. Rothfels (Munich, 1922).

—— *Clausewitz, Geist und Tat*, ed. W. M. Schering (Stuttgart, 1941).

—— *Strategie aus dem Jahr 1804 mit Zusätzen von 1808 und 1809*, ed. E. Kessel (Hamburg, 1937).

—— *Principles of War* (Harrisburg, 1942).

—— *Karl und Marie von Clausewitz, Ein Lebensbild in Briefen und Tagebuchblättern*, ed. K. Linnebach (Berlin, 1916).

—— 'Über das Leben und den Charakter von Scharnhorst', in L. von Ranke (ed.), *Historisch-politische Zeitschrift*, I (1832).

CLAUSEWITZ, CARL VON, 'Über das Fortschreiten und den Stillstand der Kriegerischen Begebenheiten', *Zeitschrift für preussische Geschichte und Landeskunde*, XV (1878), 233–40.
—— *Historische Briefe über die grossen Kriegsereignisse im Oktober 1806*, ed. J. Niemeyer (Bonn, 1977).
—— *Carl von Clausewitz, Schriften, Aufsätze, Studien, Briefe*, ed. W. Hahlweg, i. (Göttingen, 1966).
—— *The Campaign of 1812 in Russia* (London, 1843).
—— *Two Letters on Strategy*, ed. and trans. by P. Paret and D. Moran (Carlisle, 1984).
DIDEROT AND D'ALEMBERT, *Encyclopédie* (Paris and Amsterdam, 1751–80).
DUMAS, MATHIEU, *Précis des événemens militaires, ou essais historiques sur les campagnes de 1799 à 1814* (19 vols.; Paris, 1817–26).
FEUQUIÈRES, A. P. MARQUIS DE, *Memoirs Historical and Military* (2 vols.; London, 1736).
FOLARD, JEAN CHARLES DE, *Histoire de Polybe* (Amsterdam, 1774), including Folard's related military works.
FREDERICK THE GREAT, *Œuvres*, ed. J. D. E. Preuss, vols. 28–30 (Berlin, 1856).
—— *Werke*, ed. G. B. Volz (Berlin, 1913), vol. vi.
—— *Posthumous Works*, trans. T. Holcroft (13 vols.; London, 1789).
GROTIUS, HUGO, *The Rights of War and Peace* (London, 1738).
GUIBERT, JACQUES ANTOINE HIPPOLYTE DE, *A General Essay on Tactics* (2 vols.; London, 1781).
—— *Œuvres militaires* (5 vols.; Paris, 1803).
—— *Observations on the Military Establishment and Discipline of the King of Prussia* (London, 1780).
—— *Journal d'un voyage en Allemagne 1773* (2 vols.; Paris, 1803).
HEGEL, G. W. F., *The Philosophy of Right* (Oxford, 1942).
HERDER, G. J., *Outlines of a Philosophy of the History of Man* (London, 1800).
JOMINI, ANTOINE HENRI, *Treatise on Grand Military Operations* (2 vols.; New York, 1865).
—— *Summary of the Art of War* (New York, 1854; another edn. Philadelphia, 1862).
—— *Histoire critique et militaire des guerres de la révolution* (Paris, 1820–4).
—— *Introduction à l'étude des grandes combinaisons de la stratégie et de la tactique* (Paris, 1829).
JOMINI, ANTOINE HENRI, *Tableau analytique des principales combinaisons de la guerre et de leur rapports avec la politique des états* (Paris, 1830).
—— *Life of Napoleon* (4 vols.; New York, 1864).
KANT, IMMANUEL, *The Critique of Judgment* (Oxford, 1961).

LESPINASSE, MLLE J. DE, *Letters*, trans. K. P. Wormeley (London, 1902; a fuller edn. London, 1929).
LINDENAU, KARL FRIEDRICH VON, *Über die höhere preussische Taktik* (Leipzig, 1790).
LIPSIUS, JUSTUS, *Sixe Bookes of Politickes or Civil Doctrine* (London, 1594).
—— *De militia Romana* (Antwerp, 1596).
LLOYD, HENRY HUMPHREY EVANS, *The History of the Late War in Germany between the King of Prussia and the Empress of Germany and her Allies* (2 vols.; London, 1781, 1784).
—— *An Essay on the English Constitution* (London, 1770).
—— *An Essay on the Theory of Money* (London, 1771).
—— *A Rhapsody of the Present System of French Politics, of the Projected Invasion and the Means to Defeat It* (London, 1779).
—— *A Political and Military Rhapsody on the Invasion and Defence of Great Britain and Ireland* (London, 1798; a later edn. of the former entry).
LOSSAU, F. K. VON, *Der Krieg* (Leipzig, 1815).
—— *Ideale der Kriegführung* (3 vols.; Berlin, 1836).
MACHIAVELLI, NICCOLÒ, *The Chief Works and Others*, ed. Allan Gilbert (Durham, 1965).
MAIZEROY, PAUL GIDEON JOLY DE, *Cours de tactique, théoretique, pratique et historique* (4 vols.; Paris, 1766, 1785).
—— *A System of Tactics* (2 vols.; London, 1781).
—— *Théorie de la guerre* (Nancy, 1777).
MILLER, FRANZ, *Reine Taktik* (2 vols.; Stuttgart, 1787–8).
MONTECUCCOLI, RAIMONDO, *Ausgewählte Schriften des Raimund Fursten Montecuccoli*, ed. A. Veltzé (4 vols.; Vienna and Leipzig, 1899–90).
MONTESQUIEU, C. L. DE SECONDAT, *The Spirit of the Laws*, trans. T. Nugent (New York, 1949).
NAPOLEON, 'Military Maxims', in *Roots of Strategy* (Harrisburg, 1940).
—— *Mémoires pour servir à l'histoire de France sous Napoléon, écrits à Sainte Hélène*, ed. C. J. F. T. Montholon (Paris, 1823).
—— *Talks of Napoleon at St. Helena*, ed. G. Gourgaud (6 vols.; London, 1904).
—— *Notes inédites de l'Empereur Napoléon Ier sur les mémoires militaires du général Lloyd* (Bordeaux, 1901).
NICOLAI, FERDINAND FRIEDRICH VON, *Versuch eines Grundrisses zur Bildung des Offiziers* (Ulm, 1775).
—— *Essai d'architecture militaire* (n.p., 1755).
NOCKHERN VON SCHORN, FRANÇOIS, *Versuch über ein allgemeines System aller militairischen Kenntnisse* (Nuremberg, 1785).
PUYSÉGUR, J. F. C. MARQUIS DE, *Art de la guerre par principes et par règles* (2 vols.; Paris, 1748).

RÜHLE VON LILIENSTERN, J. J. A., *Handbuch für den Offizier, zur Belehrung im Frieden und Zum Gebrauch im Felde* (Berlin, 1817).
—— *Rühle von Lilienstern et son Apologie de la guerre*, ed. and trans., Louis Sauzin (Paris, 1937).
SANTA CRUZ DE MARZENADO, N. O., *Reflections, Military and Political* (London, 1737).
SAXE, MAURICE DE, *Mes rêveries* (2 vols.; Amsterdam and Leipzig, 1757).
—— *Reveries*, T. Phillips (ed.), *Roots of Strategy* (Harrisburg, 1940).
—— *Esprit des lois de la tactique* (2 vols.; La Haye, 1762).
SCHARNHORST, GERHARD VON (ed.), *Militair Bibliothek* (Hanover, 1782–4).
—— (ed.), *Bibliothek für Offiziere* (4 issues; Göttingen, 1785).
—— *Handbuch für Offiziere in den angewandten Theilen der Krieges Wissenschaften* (3 vols.; Hanover, 1787–90).
—— *Militairisches Taschenbuch zum Gebrauch im Felde* (3rd edn.; Hanover, 1794).
—— *Militärische Schriften von Scharnhorst*, ed. C. von der Goltz (Berlin, 1881).
—— *Ausgewählte Schriften*, ed. U. von Gersdorff (Osnabrück, 1983).
—— *Scharhorsts Briefe*, ed. K. Linnebach (Munich and Leipzig, 1914).
SCHLEIERMACHER, F. D. E., *On Religion* (London, 1893).
STAËL, MME LA BARONNE DE, *Œuvres complètes* (Paris, 1821), ed. by her son.
TEMPELHOFF, G. F. VON, *History of the Seven Years War* (2 vols.; concise edn: London, 1793).
TURPIN DE CRISSE, L. DE, *The Art of the War* (2 vols.; London, 1761).
VAUBAN, SÉBASTIEN LE PRESTRE DE, *Traité de l'attaque et de la défense des places* (2 vols.; La Hare, 1737).
VENTURINI, JOHANN GEORG JULIUS, *Lehrbuch der angewandten Taktik, oder eigentlichen Kriegswissenschaft* (2 vols.; Schleswig, 1800).
VOLTAIRE, 'Tactics', in *Works*, trans. T. Smollett (London, 1799–81), misc. i. 126–30.
WAGNER, A., *Grundzüge der reinen Strategie* (Amsterdam, 1809).
ZANTHIER, FRIEDRICH WILHELM VON, *Versuch über die Kunst den Krieg zu studiren* (n.p., 1775).
—— *Versuch über die Märsche der Armeen, die Läger, Schlachten und den Operations Plan* (Dresden, 1778).

Works Composed after 1837

ADAMS, F. D., *The Birth and Development of the Geological Sciences* (London, 1938).

AESCH, A. G. VON, *Natural Science in German Romanticism* (New York, 1941).

ALEXANDER, W. M., *Johann Georg Hamann, Philosophy and Faith* (The Hague, 1966).

ALGER, J. I. *Antoine Henri Jomini: A Bibliographical Survey* (West Point, 1975).

ANDRÉ, LOUIS, *Michel le Tellier et Louvois* (Paris, 1943).

ANGELI, M. E. VON, Erzherzog Carl von Oesterreich als Feldherr und Heersorganisator (Vienna and Leipzig, 1896).

ARIS, REINHOLD, *History of Political Thought in Germany from 1789 to 1815* (London, 1965).

ARON, RAYMOND, *Clausewitz, den Krieg denken* (Frankfurt am Main, 1980).

—— *Peace and War, A Theory of International Relations* (New York, 1967).

—— 'What is a Theory of International Relations?', *Journal of International Affairs*, XXI (1967), 185–206.

BAHN, RUDOLF, *Georg Heinrich von Berenhorst* (doctoral diss. Halle, 1910).

BARKER, THOMAS M., *The Military Intellectual and Battle: Montecuccoli and the Thirty Years War* (New York, 1975).

BAYLEY, C. C., *War and Society in Renaissance Florence* (Toronto, 1961).

BECK, L. W., *Early German Philosophy, Kant and his Predecessors* (Cambridge Mass., 1969).

BECKER, CARL, L., *The Heavenly City of the Eighteenth Century Philosophers* (New Haven, 1932).

BEHRENS, C. B. A., 'Which Side was Clausewitz On?', in *The New York Review of Books*, 14 Oct. 1976.

BERLIN, ISAIAH, *Against the Current: Essays in the History of Ideas* (Oxford, 1981).

—— *Vico and Herder* (London, 1976).

BLANNING, T. C. W., *Reform and Revolution in Mainz 1743–1803* (Cambridge, 1974).

BLOCK, WILLIBALD, 'Die Condottieri: Studien über die sogenannten "unblutigen Schlachten"', *Historische Studien*, CX (1913).

BLOMFIELD, R., *Sébastien Le Prestre de Vauban* (London, 1938).

BORGERHOFF, E. B. O., *The Freedom of the French Classicism* (Princeton, 1950).

BRODIE, BERNARD, *War and Politics* (New York, 1973).

BRUMFITT, J. H., *Voltaire—Historian*, (Oxford, 1958).

BRUNSCHWIG, HENRI, *Enlightenment and Romanticism in Eighteenth Century Prussia* (Chicago, 1974).

BURKE, PETER, *The Renaissance Sense of the Past* (London, 1969).

—— *Montaigne* (Oxford, 1981).

BURY, J. B., *The Idea of Progress* (New York, 1932).

BUTTERFIELD H., *The Statecraft of Machiavelli* (London, 1955).

CAEMMERER, R. VON, *The Development of Strategical Science during the 19th Century* (London, 1905).

—— *Clausewitz* (Berlin, 1910).

CAMON, H., *La Guerre Napoléonienne* (5 vols.; Paris, 1907).

—— *Le Système de guerre de Napoléon* (Paris, 1923).

—— *Clausewitz* (Paris, 1911).

CAMPORI, CESARE, *Raimondo Montecuccoli, la sua famiglia e i suoi tempi* (Florence, 1876).

CARRIAS, E., *La Pensée militaire allemande* (Paris, 1948).

—— *La Pensée militaire français* (Paris, 1960).

CASSIRER, ERNST, *The Philosophy of the Enlightenment* (Princeton, 1951).

CHANDLER, DAVID, *The Campaigns of Napoleon* (New York, 1966).

—— *The Art of Warfare in the Age of Marlborough* (London, 1976).

CHICHESTER, H. MANNERS, 'Lloyd', in *Dictionary of National Biography* (London, 1909), xi. 1301–2.

COHEN, HERMANN, *Von Kants Einfluss auf die deutsche Kultur* (Berlin, 1883).

COLIN, J., *L'Éducation militaire de Napoléon* (Paris, 1901).

—— *L'Infanterie au XVIIIe siècle* (Paris, 1907).

—— *La Tactique et la discipline dans les armées de la Révolution* (Paris, 1902).

—— *Les Campagnes du Maréchal de Saxe* (3 vols.; Paris, 1901–6).

COURVILLE, XAVIER DE, *Jomini ou le devin de Napoléon* (Paris, 1935).

CREUZINGER, PAUL, *Hegels Einfluss auf Clausewitz* (Berlin, 1911).

CREVELD, MARTIN VAN, *Supplying War* (Cambridge, 1977).

CRISTE, O., Erzherzog Carl von Oesterreich (Vienna and Leipzig, 1912).

D'ALDÉGUIER, F., *Discours sur la vie et les écrits de Guibert* (Paris, 1855).

DEBUS, A. G., *The Chemical Philosophy, Paracelsian Science and Medicine in the Sixteenth and Seventeenth Centuries* (2 vols.; New York, 1977).

DELBRÜCK, HANS, *History of the Art of War within the Framework of Political History*, vols. 1–4 (London, 1975–85).

DUFFY, CHRISTOPHER, *The Army of Frederick the Great* (London, 1974).

—— *Fire and Stone, the Science of Fortress Warfare 1660–1860* (London, 1975).

—— *Siege Warfare: The Fortress in the Early Modern World 1494–1660* (London, 1979).

—— *The Fortress in the Age of Vauban and Frederick the Great 1660–1789* (London, 1985).

EARLE, E. M. (ed.), *Makers of Modern Strategy, Military Thought from Machiavelli to Hitler* (Princeton, 1943).

EASLEA, BRIAN, *Witch-hunting, Magic and the New Philosophy* (Sussex, 1980).

ELKAN, A., 'Die Entdeckung Machiavellis in Deutschland zu Beginn des 19. Jahrhunderts', *Historische Zeitschrift*, CXIX (1919), 427–58.

ENGELL, JAMES, *The Creative Imagination, Enlightenment to Romanticism* (Cambridge Mass., 1981).

EPSTEIN, KLAUS, *The Genesis of German Conservatism* (Princeton, 1966).

EVANS, R. J. W., *Rudolf II and his World, a Study in Intellectual History 1576–1612* (Oxford, 1973).

—— *The Making of the Habsburg Monarchy 1550–1700* (Oxford, 1979).

FULLER, J. F. C., *The Foundations of the Science of War* (London, 1925).

GALLIE, W. B., *Philosophers of Peace and War, Kant, Clausewitz, Marx, Engels and Tolstoy* (Cambridge, 1978).

GAT, AZAR, 'Clausewitz on Defence and Attack', *Journal of Strategic Studies*, X, (March, 1988).

GAY, PETER, *The Enlightenment; an Interpretation* (2 vols.; London, 1967–9).

—— *Voltaire's Politics* (Princeton, 1959).

—— *Le Général Antoine Henri Jomini 1779–1869, Bibliothèque Historique Vaudoise*, 41 (Lausanne, 1969).

GIBSON, R. W., *Bacon: A Bibliography of his Works and Baconiana to the Year 1750* (Oxford, 1950).

GILBERT, FELIX, *Machiavelli and Guicciardini* (Princeton, 1965).

GILMORE, M. P., 'The Renaissance Conception of the Lessons of History', in his *Humanists and Jurists* (Cambridge Mass., 1963).

GUY, BASIL, 'The French Image of China before and after Voltaire', *Studies on Voltaire and the Eighteenth Century*, XXI, (1963).

HAGEMANN, ERNST, *Die deutsche Lehre vom Kriege, von Berenhorst zu Clausewitz* (Berlin, 1940).

HAHLWEG, WERNER, *Clausewitz, Soldat, Politiker, Denker* (Göttingen, 1957).

—— 'Philosophie und Theorie bei Clausewitz', in Clausewitz Gesellschaft (ed.), *Freiheit ohne Krieg* (Bohn, 1980), 325–32.

—— *Die Heersreform der Oranier und die Antike* (Berlin, 1941).

—— (ed.), *Klassiker der Kriegskunst* (Darmstadt, 1960).

HALE, J. R., *Renaissance War Studies* (London, 1983).

—— *War and Society in Renaissance Europe 1450–1620* (Leicester, 1985).

—— The military chapters in the *New Cambridge Modern History* (vols i–iii; 1957–8).

HANDEL, MICHAEL (ed.), *Clausewitz and Modern Strategy* (London, 1986).

HARRASCHIK-EHL, C., *Scharnhorsts Lehrer: Graf Wilhelm von Schaumburg-Lippe in Portugal* (Osnabrück, 1974).

HAUSER, ARNOLD, *The Social History of Art* (London, 1951), ii.

HAY, DENYS, *Annalists and Historians* (London, 1977).

HAZARD, PAUL, *The European Mind 1680–1715* (London, 1953).

—— *European Thought in the Eighteenth Century* (London, 1954).

HILLGARTH, J. N., *Raymon Lull and Lullism* (Oxford, 1971).

HOBOHM, WALTER, *Machiavellis Renaissance der Kriegskunst* (Berlin, 1913).

HÖHN, REINHARD, *Revolution, Heer, Kriegsbild* (Darmstadt, 1944).

—— *Scharnhorsts Vermächtnis* (Bonn, 1952).

HOWARD, MICHAEL ELIOT (ed.), *The Theory and Practice of War* (London, 1965).

—— *Clausewitz* (Oxford, 1983).

—— *War in European History* (London, 1976).

HUGHES, G. L. T., *Romantic German Literature* (London, 1979).

IGGERS, G. G., *The German Conception of History, the German Tradition of Historical Thought from Herder to the Present* (Middletonn Conn., 1968).

JÄHNS, MAX, *Geschichte der Kriegswissenschaften* (3 vols.; Munich and Leipzig, 1889–91).

JANY, KURT, *Geschichte der Königlich Preussischen Armee* (4 vols.; Berlin, 1928–37).

KAUFMANN, HANS, 'Raimondo Graf Montecuccoli, 1609–1680' (doct. diss.; Berlin, 1972).

KESSEL, EBERHARD, 'Georg Heinrich von Berenhorst', *Sachsen und Anhalt*, IX (1933), 161–98.

—— 'Zur Entstehungsgeschichte von Clausewitz Werk "Vom Kriege"', *Historische Zeitschrift*, CLII (1935), 97–100.

—— 'Carl von Clausewitz: Herkunft und Persönlichkeit', *Wissen und Wehr* XVIII (1937).

—— 'Zur Genesis der modernen Kriegslehre', *Wehrwissenschaftliche Rundschau*, III/9 (1953).

—— 'Die doppelte Art des Krieges', *Wehrwissenschaftliche Rundschau*, IV/9 (1954), 298–310.

KIERNAN, COLIN, 'Science and the Enlightenment in Eighteenth Century France', in T. Besterman (ed.), *Studies on Voltaire and the Eighteenth Century*, LIX (1968).

KLIPPEL, G. H., *Das Leben des Generals von Scharnhorst* (3 vols.; Leipzig, 1869–71).

KNUDSEN, JONATHAN, B., *Justus Möser and the German Enlightenment* (Cambridge, 1986).

KRIEGER, LEONARD, *The German Idea of Freedom, History of a Political Tradition* (Boston, 1957).

LATREILLE, A., *L'Armée et la nation à la fin de l'ancien régime* (Paris, 1914).

LAZARD, P. E., *Vauban* (Paris, 1934).

LECOMTE, FERDINAND, *Le Général Jomini, sa vie et ses écrits* (Paris and Lausanne, 1860).

LEHMANN, MAX, *Scharnhorst* (2 vols.; Leipzig, 1886–7).

LENIN, V. I., 'The Collapse of the Second International', *Collected Works* (Moscow, 1964), vol. 21.

LÉONARD, E. G., *L'Armée et ses problèmes au XVIIIe siècle* (Paris, 1958).

LIDDELL HART, B. H., *The Ghost of Napoleon* (New Haven, 1934).

—— *Strategy, the Indirect Approach* (London, 1954).

LLOYD, HANNIBAL EVANS, *Memoir of General Lloyd* (London, 1842).

LORENZ, REINHOLD, 'Erzherzog Carl als Denker', in August Faust (ed.), *Das Bild des Krieges im deutschen Denken* (Stuttgart and Berlin, 1941), 235–76.

LOVEJOY, ARTHUR, 'Herder and the Enlightenment Philosophy of History', in his *Essays in the History of Ideas* (Baltimore, 1948) 166–82.

LUDENDORFF, ERICH, *The Total War* (London, n.d.).

LYNN, JOHN A., *The Bayonets of the Republic* (Chicago, 1984).

MCCLELLAND, CHARLES, E., *State, Society and University in Germany, 1700–1914* (Cambridge, 1980).

MALLETT, MICHAEL, *Mercenaries and Their Masters: Warfare in Renaissance Italy* (London, 1974).

MARTIN, KINGSLEY, *French Liberal Thought in the Eighteenth Century* (2nd rev. edn., London, 1954).

MARWEDEL, ULRICH, *Carl von Clausewitz, Persönlichkeit und Wirkungsgeschichte seines Werkes bis 1918* (Boppard am Rhein, 1978).

MEINECKE, R. L. F., *Cosmopolitanism and the National State* (Princeton, 1970).

—— *Machiavellism, the Doctrine of Raison d'Etat and its Place in Modern History* (London, 1957).

—— *Historicism, the Rise of a New Historical Outlook*, (London, 1972).

MILLER, R. D., *Schiller and the Ideal of Freedom, A Study of Schiller's Philosophical Works with Chapters on Kant* (Oxford, 1970).

NISBET, H. B., *Herder and the Philosophy and History of Science* (Cambridge, 1970).

NOHN, ERNST AUGUST, 'Der unzeitgemässe Clausewitz', *Wehrwissenschaftliche Rundschau*, V (1956).

OESTREICH, GERHARD, *Neostoicism and the Early Modern State* (Cambridge, 1982).

OLDRIDGE, O. A., 'Ancients and Moderns in the Eighteenth Century', in P. Wiener (ed.), *Dictionary of the History of Ideas, Studies of Selective Pivotal Ideas* (New York, 1968), i. 76–87.

OMAN, CHARLES, *The Art of War in the Middle Ages* (2 vols.; London, 1924).

—— *The Art of War in the Sixteenth Century* (London, 1937).

OMMEN, HEINRICH, 'Die Kriegsführung des Erzherzogs Carl', *Historische Studien*, XVI (Berlin, 1900).

PAGEL, WALTER, *Joan Baptista van Helmont* (Cambridge, 1982).

PARET, PETER, *Clausewitz and the State* (Oxford, 1976).

PARET, PETER, *Yorck and the Era of the Prussian Reform* (Princeton, 1966).
—— 'Education, Politics and War in the Life of Clausewitz', *Journal of the History of Ideas*, XXIX (1968), 394–408.
—— 'Die politischen Ansichten von Clausewitz', in Clausewitz Gesellschaft (ed.), *Freiheit ohne Krieg*, 332–48.
—— (ed.), *Makers of Modern Strategy from Machiavelli to the Nuclear Age* (Princeton, 1986).
PARKER, GEOFFREY, 'The "Military Revolution 1560–1660"—a Myth?' *Journal of Modern History*, XXXXVIII (1976), 195–214.
PARKINSON, ROGER, *Clausewitz* (London, 1970).
PASCAL, ROY, *The German Sturm und Drang* (London, 1953).
PEBALL, KURT, 'Zum Kriegsbild der österreichischen Armee und seiner geschichtlichen Bedeutung in den Kriegen gegen die Französische Revolution und Napoleon I', in Groote, W. V., and Müller, K. J. (eds), *Napoleon I und das Militärwesen seiner Zeit* (Freiburg, 1968), 129–82.
PERIÉS, G., 'Army Provisioning, Logistics and Strategy in the Second Half of the 17th century', *Acta Historica Hungaricae* XVI 1–2 (1970), 1–51.
PICHAT, HENRY, *La Campagne du Maréchal de Saxe 1745–6* (Paris, 1909).
PIERI, PIERO, 'La formazione dottrinale di Raimondo Montecuccoli', *Revue internationale d'histoire militaire*, III (1951), 92–125.
PINSON, K. S., *Pietism as a Factor in the Rise of German Nationalism* (New York, 1968).
PÖHLER, J., *Bibliotheca historico-militaris* (4 vols.; Leipzig, 1887–97).
POIRIER, LUCIEN, *Les Voix de la stratégie—Guibert* (Paris, 1977).
POTEN, K. G. H. B. VON, *Geschichte des militär-Erziehungs und Bildungswesens in den Landen deutscher Zunge*, vols. 10, 11, 15, 17, 18 of C. Kehrbach (ed.), *Monumenta Germaniae Pedagogica* (Berlin, 1889–97).
PRIESDORFF, KURT VON, *Soldatisches Führertum* (10 vols.; Hamburg, 1936–42).
QUIMBY, ROBERT, *The Background of Napoleonic Warfare, the Theory of Military Tactics in Eighteenth Century France* (New York, 1957).
RANDALL Jun., J. H., *The Career of Philosophy from the Middle Ages to the Enlightenment* (New York, 1962).
REIL, PETER HANNS, *The German Enlightenment and the Rise of Historicism* (Berkeley, 1975).
RITTER, GERHARD, *The Sword and the Scepter* (Miami, 1969).
ROBERTS, MICHAEL, *Gustavus Adolphus* (2 vols.; London, 1958).
—— 'The Military Revolution 1560–1660' in his *Essays in Swedish History* (London, 1967).
ROSINSKI, HERBERT, 'Die Entwicklung von Clausewitz Werk "Vom Kriege" im Licht seiner "Vorreden" und "Nachrichten"', *Historische Zeitschrift*, CLI (1935), 278–93.

ROTHENBERG, G. E., *Napoleon's Great Adversaries, The Archduke Charles and the Austrian Army, 1792–1814* (London, 1982).

—— KIRÁLY, B. K. AND SUGAR, P. F. (eds), *East Central European Society and War in the Pre-Revolutionary Eighteenth Century*, vol. ii of *War and Society in East-Central Europe* (New York, 1982).

ROTHFELS, HANS, *Carl von Clausewitz, Politik und Krieg* (Berlin, 1920).

SAINTE-BEUVE, C. A., *Le Général Jomini, étude* (Paris, 1869).

SANDONNINI, TOMMASO, *Il Generale Raimondo Montecuccoli e la sua famiglia* (Modena, 1914).

SAUNDERS, J. L., *Justus Lipsius, the Philosophy of Renaissance Stoicism* (New York, 1955).

SCHENK, H. G., *The Mind of the European Romantics* (London, 1966).

SCHERING, WALTER MALMSTEN, *Die Kriegsphilosophie von Clausewitz* (Hamburg, 1935).

—— *Wehrphilosophie* (Leipzig, 1939).

SCHOTTEN, COL. J. VON, *Was muss ein Offizier wissen?* (Dessau and Leipzig, 1782).

SCHRAMM, WILHELM VON, *Clausewitz, Leben und Werk* (Munich, 1976).

SCHWARTZ, KARL, *Leben des Generals Carl von Clausewitz und der Frau Marie von Clausewitz* (2 vols.; Berlin, 1878).

SHANAHAN, WILLIAM, *Prussian Military Reforms 1786–1813* (New York, 1945).

SHARPE, LESLEY, *Schiller and the Historical Character* (Oxford, 1962).

SIMON, W. M., *Friedrich Schiller: the Poet as Historian* (Keele, 1966).

SKINNER, QUENTIN, *The Foundations of Modern Political Thought* (Cambridge, 1978).

STADELMANN, RUDOLF, *Scharnhorst, Schicksal und geistige Welt, ein Fragment*, (Wiesbaden, 1952).

STERN, J. P., *Lichtenberg, a Doctrine of Scattered Occasions* (Indiana, 1959).

STRAUSZ-HUPÉ, ROBERT, *Geopolitics: The Struggle for Space and Power* (New York, 1942).

TAYLOR, F. L., *The Art of War in Italy 1494–1529* (Cambridge, 1921).

THORNDIKE, LYNN, *A History of Magic and Experimental Science* (New York, 1923–58) vols. vi–viii.

TOOLEY, R. V. AND BRICKER, C., *A History of Cartography* (London, 1968).

TROELTSCH, ERNST, 'The Idea of Natural Law and Humanity in World Politics', Appendix to O. Gierke, *Natural Law and the Theory of Society 1500–1800* (Cambridge, 1934), 201–22.

TYMMS, RALPH, *German Romantic Literature* (London, 1955).

USCZECK, HANSJÜRGEN, *Scharnhorst, Theoretiker, Reformer, Patriot* (East Berlin, 1979).

VAGTS, ALFRED, *A History of Militarism* (London, 1959).

VEHSE, E., *Memoirs of the Court and Aristocracy of Austria* (2 vols.; London, 1856).

VENTURI, FRANCO, 'Le avventure del generale Henry Lloyd', *Rivista Storia Italiana* [RSI], XCI (1979), 369–433.

WADE, IRA O., *The Structure and Form of the French Enlightenment* (2 vols.; Princeton, 1977).

—— *The Intellectual Origins of the Enlightenment* (Princeton, 1971).

WAGNER, GEORGE, *Das Türkenjahr 1664, Raimund Montecuccoli, die Schlacht von St. Gotthard-Mogarsdorf*, issue 48 (1964) of *Burgenländische Forschungen*.

WAGNER, WILHELM, *Die preussischen Reformer und die zeitgenössische Philosophie* (Cologne, 1956).

WALKER, D. P., *Spiritual and Demonic Magic from Ficino to Campanella* (London, 1958).

WALLACH, JEHUDA L., *Das Dogma der Vernichtungsschlacht* (Frankfurt am Main, 1967).

—— *Kriegstheorien, ihre Entwicklung im 19. und 20. Jahrhundert* (Frankfurt am Main, 1972).

WARD, ALBERT, *Book Production, Fiction and the German Reading Public, 1740–1800* (Oxford, 1974).

WELLEK, RENE, *A History of Modern Criticism* (London, 1955), vol. i.

WELLS, G. A., *Herder and After* (Gravenhage, 1959).

—— *Goethe and the Development of Science* (The Netherlands, 1978).

—— 'Herder's Two Philosophies of History', *Journal of the History of Ideas*, XXI (1960), 527–37.

WENIGER, ERICH, 'Philosophie und Bildung im Denken von Clausewitz', in W. Hubatsch (ed.), *Schicksalswege Deutscher Vergangenheit* (Düsseldorf, 1950), 123–43.

WILKINSON, SPENSER, *The French Army before Napoleon* (Oxford, 1915).

WILLISEN, WILHELM VON, *Theorie des grossen Krieges* (Berlin, 1840).

WRIGHT, C. H. C., *French Classicism* (Cambridge Mass., 1920).

YATES, F. A., *Giordano Bruno and the Hermetic Tradition* (London, 1964).

—— *The Art of Memory* (London, 1966).

—— *Theatre of the World* (London, 1969).

—— *The Occult Philosophy in the Elizabethan Age* (London, 1979).

YORK VON WARTENBURG, *Napoleon as a General* (2 vols.; London, 1902).

ZEISSBERG, H. R. VON, *Erzherzog Carl von Oesterreich* (Vienna and Leipzig, 1895).

INDEX